UNIVERSITY LIBRARY
NEWCASTLE UPON TYNE

WITHDRAWN

Continuous Quality Improvement

NEWCASTLE UNIVERSITY LIBRARY
097 09629 5
658.562 WHI
LONG

Also by Alasdair White:

Managing For Performance – How to get the best out of yourself and your team (Piatkus)

CONTINUOUS QUALITY IMPROVEMENT

A HANDS-ON GUIDE TO SETTING UP & SUSTAINING A COST-EFFECTIVE QUALITY PROGRAMME

ALASDAIR WHITE

Copyright: © 1996 Performance Management Solutions Ltd

All rights reserved

First published in 1996 by
Judy Piatkus (Publishers) Ltd,
5 Windmill Street, London W1P 1HF

The moral right of the author has been asserted

A catalogue record of this book is available from the British Library

ISBN 0 7499 1570–6 (hbk)

Edited by Betty Palmer
Designed by Chris Warner

Data capture & manipulation by
Action Typesetting Limited, Gloucester
Printed & bound in Great Britain by
Mackays of Chatham PLC

For my father, who
started me on my journey

Acknowledgements

As with any major undertaking, writing and publishing a book is not a solitary process. The actual responsibility for putting the words on paper was mine and any errors or omissions are also my responsibility, but many others have contributed to the final outcome. My thanks, therefore, to: Karl Sergeant, who first taught me about the quality movement; Philippe Paillart, who allowed me the space to develop the CQI process; Citibankers throughout Europe, who, wittingly or unwittingly, tested the ideas and provided feedback on them; John Hughes-Wilson, who carried out the 'sanity checks' whenever I required them; Sarah Hannigan and the Piatkus team, who publish my work; and lastly my wife Fiona, who patiently checked the manuscript and helped in the proofreading.

Alasdair White
Belgium

Contents

Introduction

Over the years, various management theories have claimed that they are the panacea to cure all ills in business. In reality all of them contributed to the greater understanding of the process of managing a company and none of them was a complete answer in itself. Recently, three of the most powerful ideas have focused on mathematical modelling, re-designing the business processes and quality. None of the theories put forward by the 'gurus', with the notable exception of Peter Drucker, has focused on that most essential of activities – the management and leadership of people as the core management function. Get this right and just about everything else will look after itself.

People – the most valuable and expensive of resources. People – frequently the most misused of resources. People – the most misunderstood and difficult to control of resources.

I think we should be spending much more time managing our people. If we did, then we would obtain better performances from them, higher returns on our investment in them, and vastly enhanced profitability for the company. What is needed now is for companies to re-focus on the correct management of people. We must put away the management tools of modelling, re-design, financial engineering, automation and the like and only take them out when needed for a specific job. We must go back to basics and manage people for performance.

We must also work towards continuous improvement in all aspects of our business. We need an overall strategy that co-ordinates and links the company's activities and focuses them on a continuous improvement in the quality of every aspect of the company. Companies that recognise and implement such a strategy – and move beyond just paying lip-service to it – will be the outstanding businesses of the future.

Continuous Quality Improvement is just such a strategy. It is a people-based programme focusing on the better management of our human resources while, at the same time, seeking improvements in all the functions and processes within the business. It is cost-effective: it need not involve large budgets but the investment returns are enormous. One client of mine achieved a 40% increase in revenue (some $10 million) by the committed application of CQI and the performance management techniques involved.

Achieving continuous quality improvement is vital. There is not one single company today that can afford to ignore the need to improve the way it does business.

In this book, I have outlined a comprehensive programme for achieving CQI, a programme that is straightforward but highly focused and based on proven ideas and techniques. By embarking on CQI you will be starting a journey that has no end but is worthwhile and rewarding.

Part I

A New Strategy

1

The Need for a New Approach

Companies face challenge • survival depends on evolution • quality programmes – a step in the right direction • THE business survival strategy

IT IS AN ALARMING THOUGHT, but the vast majority – over 75% – of all companies in existence today will not be around in 60 years' time. They will have been taken over, merged, bought out, gone bankrupt, or gone into liquidation. Those that do survive will only do so in a radically different form.

In the modern business world, the average life expectancy of a company is 60 years, assuming it survives the first two years of existence. Many do not outstay the working life of their founders and a very large number do not last that long. Of the original 12 companies that made up the Dow Jones Stock Market Index in New York in 1912, only two still survive under their original names. A check on the FT Index of the London Stock Exchange shows a similar result. If you look at the leading businesses in almost any country, they are either still under the control of the founder or original directors or are the result of recent acquisitions, mergers, or takeovers – in each case, the current form of the company is generally less than 25 years old and often less than 10 years old.

So why do companies change?

COMPANIES FACE CHALLENGES

In simple terms, they change so that they can meet the challenges facing them in their marketplace. The challenges facing most businesses in today's marketplace are

- shorter product cycles
- products and services rapidly become generic
- increased competition for fixed or diminishing markets
- higher customer expectations
- higher staff expectations

Let's take a look at these in greater detail.

Shorter product cycles

It must seem to many managers that the product cycle from conception through development, launch and marketing to loss of market share and eventual obsolescence is becoming ever shorter. The Japanese, in particular, recognise this and are for ever seeking improvements, modifications, and changes in image for their products, especially in the consumer field. You only have to think of the Sony Walkman to see this in action. The original Walkman and the modern version do exactly the same thing, but the one I recently bought for my children looks very different from the early model I have for my own use – and it cost less than I originally paid for mine, never mind inflation.

Although the basic function of many products and services stays the same over a very long period of time, the market for them diminishes unless they change to meet changing fashions, fads and other market-driven requirements. A classic example of this is the motor car: unchanged in basic function for over 90 years – but, oh! how different in design, comfort and performance. Most car manufacturers have spent millions of pounds, dollars, D-marks, and yen to make their product more attractive to the customer and yet one of the most common phrases heard about models is that they have become 'dated'.

Product and services rapidly become generic

Just when we think we have a market leader with our new product or service, along comes a competitor with the same thing – occasionally better and frequently cheaper. The ability of our company to stay ahead in the marketplace is often based on our skill at developing products that are at the leading edge of technology – and staying one step ahead of the competition. This requires investment in Research & Development and an active and enterprising marketing approach to sell the new ideas to the customer. But, despite all our efforts, our competitors are never far behind and our brand new market-leading product is now available from them as well and we've lost our advantage – our products have become generic; everybody sells one.

Increased competition for fixed or diminishing markets

Many of the traditional markets have reached maturity and are no longer growing – some have even contracted – and yet companies are now facing increased competition as more and more businesses enter the field in search of market share and increased profits. Some respond by taking over competitors to increase their market share, while others alter their approach and become niche businesses. In either case, the result is a significant change within the company – or it should be.

Higher customer expectations

As markets mature, customers become more sophisticated about their requirements from the goods and services on offer in that market. No longer will they put up with less than top quality. In a market of more or less fixed size and with a widening range of suppliers, customers reign supreme: they can choose to be choosy and they can demand higher and higher levels of service and quality. Mature markets are buyers' markets and if a company is to survive it must listen to the customers and work to meet their expectations.

The Henry Ford approach to marketing – any colour you like as long as it's black – no longer works. Companies aiming to survive are having to move to just-in-time production with special orders and customised solutions being delivered as quickly as standardised ones, if not quicker. A friend of mine has just bought a new car from a European manufacturer and experienced this at first hand: he could have a standard version of the car he wanted in a predetermined colour within a week, but he wanted a specific selection of optional extras, including an optional colour. The car, as ordered, was delivered to him in five working days – an impressive display of personal service.

Meeting higher customer expectations requires a major change in the way we do business and for many companies this means they have to reinvent themselves regularly.

Higher staff expectations

Even as recently as fifteen years ago many workers believed that their job was theirs for life. They expected to receive more pay each year, perhaps a shorter working week, and increased holidays as they achieved seniority. Promotion was based on seniority and there was generally a clear career path to the top. People expected to stay with the same company for the greater part of their working lives; some still make this assumption.

But changes have occurred – some brought about by the current business thinking of 'cut costs to improve profit' which inevitably results in staff losses, and some brought about by a change in people's attitudes towards work.

No longer do people believe in a job for life – they know that at the first sight of a downturn in the market their job will be on the line. Now their attitude is more cynical, many employees are constantly on the lookout for a better job in a different company. Flatter organisations tend to have the same effect – where job promotion (vertical movement in a hierarchy) is not possible promotion must come by changing jobs, either within a company or by moving to a different employer.

In some industries, notably in computer programming and sales functions, it is an accepted part of human resource management that people will stay, on average, three to five years and it is almost the norm that staff at almost any level will have had between five and eight different career moves in their working life. Such job changes have, in the past, been seen as a weakness, raising questions about a person's reliability and commitment, but now they are accepted as indicating a desire for career and individual development.

Given this greater flexibility in the attitudes of the working population, companies are having to make changes in the way they manage and reward their staff. Unfortunately, this is one area in which change is slow and it really does need to be at the top of every manager's agenda if the right people are to be attracted and retained.

SURVIVAL DEPENDS ON EVOLUTION

Clearly, companies must change if they are to survive and prosper. They must evolve, become more people-orientated, and be more conscious of the expectations of their customers and staff. They must go back to the basics of good management but on a much higher plane. If the change is not continuous it is not evolution, and if the company is not evolving then it will die in just the same way as Darwin deduced from his observations in the Galapagos Islands. Evolution, in business terms, is a matter of adapting to the change in the environment – it is also the 'survival of the fittest' in which only those who adapt to changes in the environment will remain fit and capable of survival.

But, if evolution is the key to survival, what happens to the highly structured and rarefied world of business modelling, forecast responses to market forces, and predictability in the marketplace? They too must evolve if they are to remain of any use at all in tomorrow's business environment.

Personally, I find much of the academic approach to business planning and corporate strategy to be of little use in the real world – a world in which people are still the key factor. Business modelling, be it in market strategy, organisational structure, or the workings of the market, relies on predictability of response – if you change one factor in the model then the outcome is predictable because of the mathematical structure of the model itself.

But people do not behave with mathematical predictability. Their response is non-rational and highly complex – more akin to the mathematics and theory of chaos – and we need a much more advanced understanding if we are to have any success in business modelling involving people. I believe that, in the fast-changing business environment, business modelling has its part to play – but only as a bit player and not in the lead role. Any company that continues to believe in the virtual infallibility of its business models risks ending up in the gutter wondering what hit it. However, those that use the models wisely to aid the planning process will do better, providing they remember to factor in the non-rational 'chaos' response of those unpredictable humans – their staff and customers.

It is a sad fact, but one that is easy to understand, that business schools and the academic world in general are hooked on business modelling with its emphasis on rigorous, supposedly scientific analysis. Business management, in fact, is much more an art than a science. Artists, however, are perceived as seldom respectable and for many business theorists the need for the respectability of the scientific or mathematical approach seduces them away from the real world. Nowhere is this more apparent than in the MBA courses so beloved of the number-loving managers of the 1970s and 80s. One colleague of mine has recently attended an MBA course as a mature student and reports that less than 15% of the course was on skills and techniques of managing people while the other 85% was on the theories and practices of accountancy, business modelling, numbers management, and other pseudo-scientific business activities. Obviously, such activities are of considerable value, but to assume that they are four times as important as managing people is ridiculous.

Quality Programmes – A Step in the Right Direction

The balance between the technical skills of management and people management skills needs redressing and the quality movement with its many different programmes and approaches is a step in the right direction.

Originating in the USA during the Cold War arms race, quality certification or registration was seen as a way to ensure that the multitude of suppliers involved in modern arms manufacture provided products that were of the right standard without the end user having to check everything. Eventually quality registration became a prerequisite for doing business with the Pentagon – although it has to be said that it did not stop a great deal of poor quality and overpriced products being supplied to the American military.

Since then there have been many advances in quality programmes – some have focused on customer service, some on production, and, more recently, some on the management process itself. As far as I am concerned, any improvement in the quality of the products, services, and management of companies is welcome and should be actively encouraged. The arrival of BSI 5750, ISO 9000 (9001–9004), and other quality registrations such as the British initiative *Investors in People* marks a move towards a better business world, but they should be seen as milestones along the way rather than an end in themselves. Unfortunately, they are often the sole desired outcome of a company's quality programme and are seen as a marketing tactic to establish a competitive advantage. Indeed, I know of a number of such programmes that have been funded from the marketing budget of the company concerned as a one-off marketing expense.

Such programmes frequently fail to deliver – especially in terms of increased revenue and profits (which is, after all, the prime purpose of being in business in the first place) – and this raises the question: why?

The most frequently quoted reasons for failure are:

- ***they are imposed from above:*** Most quality initiatives are instigated by senior management and are seen by those below as 'yet another bright idea by those who don't have to make it work'. This leads directly to:
- ***a lack of 'ownership':*** The programmes will only work if the people involved feel involved enough to take ownership of the process and make it work. Imposed programmes are seldom owned by those who are involved and this lack of commitment cannot be overcome by issuing orders from the top. Poor communication, a common business problem, is another contributory factor in lack of ownership and so is:
- ***a lack of clear objectives for the programme:*** The most frequent comment I hear on this is: 'but what are the real objectives? why are we really doing this?' For the programme to be successful we need to ensure that the objectives are clearly communicated, are thoroughly understood, and are accepted and adopted by the staff. If this is not done then the programme is seen as:

- ***frightening and threatening to the status quo:*** Job security is seen as under threat, insecurity rises, productivity drops, and the very problems the programme is supposed to address are magnified by the programme itself. This leads directly to:
- ***inertia and a resistance to change:*** Resistance to change is a huge stumbling block to progress of any sort. Once entrenched, it leads to inertia and forward movement can become impossible. Once the programme has reached this stage it is generally abandoned, thus proving the sceptics right and forcing the business to write off the expense involved. This then 'proves' that:
- ***quality programmes are expensive.*** All because they are not properly thought through, planned, and implemented.

But what happens if the programme is handled correctly? Then it becomes a strategy for survival – it becomes the driving force for the evolution of the company.

QUALITY AS THE BUSINESS SURVIVAL STRATEGY

Given that all companies, if they are going to survive, need to evolve, to reinvent themselves continuously, a strategy is needed that will ensure the evolution does not stop. But it must be the right sort of strategy, since survival is dependent on **continuously improving** the way the company operates – in terms of its processes, products and services **and** in the way it manages its most volatile and valuable resource – its people.

This strategy must be practical, it must be realistic, it must be cost-effective, and above all, it must be accepted by our people – they must buy into the strategy, they must believe in it, and they must be committed to it. In fact, they must own it. The strategy must also form the core around which all the other key business strategies operate and against which they are judged.

I call this strategy **Continuous Quality Improvement** or CQI and we will be looking at how it can be developed, what tools we need to use, the tactics involved in implementing it, and how we can ensure that it is a continuous evolution and not a one-off event. *While CQI can never be the only strategy we need,* it is **the** key to our survival. It needs to be the umbrella under which all other strategies and philosophies operate.

SUMMARY POINTS

- The average life expectancy of a company is 60 years.
- Companies must change and adapt regularly if they are to meet the challenges facing them.
- The principal challenges are
    - shorter product cycles
    - generic products
    - increased competition for fixed or diminishing markets
    - higher customer expectations
    - higher staff expectations.
- Survival depends on evolution – constant change.
- Highly structured, pseudo-scientific, quantified management processes must give way to much better management of human resources.
- Business modelling cannot handle the non-rational 'chaotic' response of staff and customers. Business models have a role to play but it is secondary to good basic people management skills.
- Quality programmes are a step in the right direction in

redressing the balance between the scientific and mathematical approach and people management.

- Quality registrations such as the ISO 9000 series are a milestone along the way and not an end in themselves.
- Quality programmes based on a fixed end objective (ISO 9000) often fail to deliver where it really matters – long term profit.
- The reasons for failure of quality programmes include
  - they are imposed from above
  - they lack ownership by the staff
  - there are no clearly communicated objectives
  - they are frightening and threatening to the status quo
  - there is resistance to change
  - they are seen as expensive
- Evolution is necessary for survival and that means continuous improvement in the way the company operates and manages.
- A strategy for continuous improvement is required – a strategy for survival.
- Continuous Quality Improvement is the key business survival strategy.

2

Continuous Quality Improvement

What is CQI • the business survival strategy • the CQI programme

OVER THE YEARS a whole school of thought has grown up over the management and development of companies which has given rise to an almost scientific theory of management. The overall problem with this approach is that it seduces managers away from the basics – the management of people. My belief is that our job as managers is to manage processes and lead people and by supplying that leadership we will get our people to achieve the best possible performance. Unfortunately, this has been hampered by the addiction of many managers to 'the numbers' and the process for delivering them.

If we are to survive, if our companies are to survive and forge ahead, we must use quality management practices to obtain top quality performance. This will mean implementing performance management – the subject of my book *Managing for Performance* – but it also entails ensuring that our company is not getting in the way of our people as they try to deliver a quality performance.

'But,' I hear you say, 'isn't that what total quality management (TQM) is about?' Yes, it is, but most TQM approaches suffer from the problems identified in Chapter One. At the same time, they are, by definition, concerned with management and we need to move away from purely management focused programmes and start using a process that actively involves everyone in improving the quality of the company. And by everyone, I mean managers of all levels, the staff, and even the customers.

Radical? Not really, just applied common sense and a clear idea of what has to be achieved. If the ultimate objective is continuous improvement, then we need to establish a central core process and procedure within the company that supports all the other activities and ensures they contribute to the quality of the company.

Continuous Quality Improvement Defined

First, let's define CQI: 'Continuous Quality Improvement is a global approach to business development that establishes an integrated programme through which a company can achieve continuous incremental improvements in its chosen key performance measures by focusing on the better leadership of people and the improved management of business processes.' This definition contains a number of key phrases that should be explored further.

Integrated programme

The first key phrase is 'integrated programme'. CQI is a process that uses the tools of the quality movement, such as TQM, re-engineering, customer care and empowerment, as and when necessary to achieve the objectives of the programme. It is a new approach to an old problem and establishes a core programme for the development of the business.

Continuous incremental improvements

CQI is not about orders-of-magnitude improvement – for that we would have to turn to process re-engineering in a big way. It is about incremental improvements in our chosen measures of performance on a continuous basis of, say, 10–20% per annum.

Key performance measures

Each company will have its own key measures of performance and most, I suggest, will be using financial criteria, but the evidence goes well beyond that and we must include the 'soft' areas as well. My experience has been that if performance in the following areas is improved, then there will be an overall improvement in the quality of the company and in the profit line:

- Human Resource Management
- Internal Service Quality
- External Service Quality
- Customer Focus
- Business Development
- Internal Relationships
- Key Financial Ratios (Net Revenue per Full Time Employee, Return on Invested Capital, Dividends to Shareholders, etc.)

Each one of these dimensions can and should be broken down into secondary performance areas and we will be looking at this in greater detail later.

Better leadership of people

Better leadership is the key to improved quality and performance within our company. The CQI programme focuses on managing for performance and all that entails in terms of human resource management.

Management of business processes

Although the CQI programme recognises that there may

be genuine reasons for process re-engineering and, indeed, encourages close scrutiny of all business processes, its primary aim is to ensure that our current procedures and systems are effective and to seek additional incremental improvements without radical change.

At the risk of being simplistic, the CQI programme is a practical and cost-effective strategy that anyone can apply within their company. It is a strategy that pulls together all the disparate parts of our business and gives it cohesion by providing a method whereby we can call on the combined strengths and knowledge of our staff to deliver our corporate goals.

THE BUSINESS SURVIVAL STRATEGY

The core to the CQI approach is the business survival strategy that can be summed up as follows:

- determine the right functions that need to be carried out
- make sure the right functions use the right processes
- get the right people to do the right jobs
- manage in the right way
- have the right customer-focused products and services – for internal as well as external customers.

All these things are important and if they are done, then the 'numbers' will take care of themselves. Given this level of importance, we need to take a look at each of them to gain an understanding of what they are and how they are interlinked. We also need to understand the opportunities and limitations that each provides. In thinking about these issues we are also involved in looking at the fundamentals of the business and some people find such a return to basics very difficult.

One client of mine insisted that he knew exactly what business he was in and of course he knew what had to be done. The trouble was he thought he was a petrol retailer with a chain of petrol stations whereas the majority of his profits came from the food, other grocery items, and the wide range of non-car-related products he sold throughout his forecourt shops. When we talked to his customers they told us they regarded his petrol stations as a replacement for the corner shop: each was, in fact, a multi-purpose, 24-hour retail shop that happened to sell petrol as well. Once the mind-set of the client had been changed we were able to focus on the real business and to develop it accordingly. He still sells petrol, but he now regards it purely as one of his retail items.

Determine the functions that need to be done

That may sound obvious, but for most companies it is actually very difficult to do. Any company more than a couple of years old has so many jobs and functions that 'need doing' that it is no longer certain what is necessary. Care must be taken to remember that a function (a business activity) and a job (an activity or role carried out by a person) are not always the same thing.

Ideally, the best starting point would be to take a clean sheet of paper and design all the processes of the company so that only the necessary functions are carried out. Unfortunately, although this may be desirable, it is hardly practical. It should certainly not be carried out at this stage – such a re-design should only be undertaken if found to be really necessary. This is something I will explore in a later chapter.

In reality most functions within a company are, to a greater or lesser extent, necessary, although it is questionable whether the company needs a specific person to carry out that function. For example: the functions of a typing pool – typing, filing, – do not really require a specific person to be employed. Many large firms do away with this job and assign the functions to managers and others who prepare their own letters directly on the word processor and file them electronically. Having orders processed, checked, entered on the client's account, and so on, are also examples of functions that no longer need to stand alone as jobs – one person entering the data in the computer can cause all these activities to take place.

Some jobs are obviously necessary – or so they seem. For example, if you are running a bank then you will need bank tellers (cashiers) – or do you? Some banks have decided that the *function* is necessary (the disbursing of cash), but the *job* (a human being doing it) is not and they have allocated the *function* to a machine. At this stage, however, we are primarily concerned with whether the function is necessary rather than whether we should employ a person or a machine to do it.

Another example might be that of salespeople. Clearly, for a business to survive it has to sell its products and services and, therefore, the *function* of sales has to be carried out. The questions you should ask are: does it have to be done by a person? Could it be done by mail order? Could it be done by a machine? Would it be done better by a person than by a machine?

In this last example we touch on a key issue: is the function connected to one of the basic essential activities of our business? All businesses have certain essential core activities – production (making the product or operating the service), marketing and sales (exchanging the product or service for the customer's money), internal administration (paying suppliers and accounting for the expenditure and income) – depending on what business we are in. Which takes us back to the earlier question: do we really know what business we are in?

The right processes for the right functions

Once we know what the necessary functions are within the business, then we can look at whether they are using the right processes. This is a more extensive subject than you might imagine and one that tends to cause managers to embark prematurely on process analysis and re-engineering. Indeed, this is exactly where we should be going but let's not rush it. The decision to go for re-engineering is one that must only be taken following rigorous analysis of the current process and a conscious decision that the benefits of the re-engineering are worth the upheaval.

I am not, at this stage, particularly interested in redesigning the whole business – I want to understand the processes involved in each function. Over the longer term we must examine each and every process to ensure that it is the right one for that function. For the moment, let us just understand that the process used in a function is the way that function is carried out and in the business survival strategy we need to ensure that we use the right processes for each function.

Get the right people to do the right jobs

Assuming that the function and its associated process involves people, we need to ensure that we have the right people in the right jobs. We also need to ensure that we have the right number of people involved in the process.

This is where we start to think about manning levels given current and future objectives. It is also where we have to consider what sort of person should be doing the job, what skills and competencies they need to meet the Minimum Performance Standards of the role – in fact, what is the profile of the person we need in the job. Naturally, this leads to a consideration of those currently doing the job – do they meet the profile, have they the skills and competencies necessary, do we have enough of them?

It is very possible that at any given time we may have people doing the wrong job: they fail to meet the profile and may be better matched to the profile of a different job. Again I stress we are not looking for cost savings, we are looking to ensure that the right people are doing the right job. Only if we do this will we be able to get the best performance from each person and the best result for the company.

Let me be quite clear. *It is unlikely that your company exists to provide employment as a social service to the community,* but it does have a responsibility to use the people it does employ in a wise and productive manner. A hire-and-fire approach is counter-productive to long-term profit and is generally a sign of weakness in the strategic thinking of senior management. That is not to say that we shouldn't take the right people on when we have a vacancy in a necessary function, nor that we shouldn't release them when that function is no longer required. What we must do, however, is plan our people requirements carefully so that we make the best use of the available resources. Having recruited someone to a function (an expensive and time consuming task in itself), if that function then becomes redundant through a change in corporate strategy we should seek to move that person to a different function rather than get rid of them (this may also be the most cost-effective solution). People are actually very flexible and a company that seeks to employ and retain its people in productive activity will generate loyalty which is a major asset to any company. Staff are genuinely interested in the success of their enterprise and have much to lose if it fails: they will, therefore, work hard to ensure that success and not failure is the outcome of their activity. We should make use of this characteristic when planning our human resource strategy.

Manage in the right way

A little earlier I defined a manager as someone who manages a process and leads people, and to manage in the right way we need to understand what this means.

Managing is the activity by which we ensure something happens according to a predetermined process and schedule – it is a function of the need for the process to be carried out in a certain way. Take a mail shot marketing campaign as an example; once the campaign has been decided upon the management process may go something like this:

- arrange development of the letter content
- determine the content of the mailshot (letter, flyer, response coupon etc.)
- determine follow-up process
- assemble mailing list and arrange for address labels
- order envelopes, flyers etc.
- check artwork
- order printing of letter and response coupon
- confirm delivery dates of printing, flyers, addressed envelopes
- organise 'stuffing' process
- receive all items and check for completion
- 'stuff' and mail
- receive responses and start follow-up

and so on.

The manager's job is to ensure that each step in the process is carried out to the agreed standard and within the agreed time frame (and, almost certainly, within the approved budget). At no time is the role other than a supervisory one and it is mechanistic. In fact, any responsible member of staff empowered with the right level of authority could carry it out. Which raises the question: why do we need a manager to do this?

The answer is, in many cases, we don't. *The supervisory role of the manager is frequently historical and based on the lack of trust many companies felt towards their people.* Fortunately, this attitude is changing. Companies are recognising that properly selected and well-trained staff can be empowered with specific authorities relating to their functions, resulting in the elimination of the need for the supervisory (managerial) role. This, in turn, has led to the wholesale elimination of managerial levels – in the down-sizing, right-sizing, de-layering 1980s and 90s it is the management grades and jobs that have gone as companies accept that they can trust their people and no longer need overseers.

Does this mean that we don't need managers? If we define the word 'manager' as meaning a supervisor, then the answer has to be: no, we don't need managers – they are an overhead without adding value. But this response changes when we consider the manager as a leader of people.

People cannot be managed in the same way as a process is managed. They have to be led, motivated, encouraged; their performance has to be monitored and feedback given. This requires different skills from those deployed in the management of a process. No longer are functional skills acceptable; high-level interpersonal skills are needed, combined with a sound knowledge of how to handle a resource whose responses cannot always be foreseen. Accurate prediction is the basis of functional management – if you do *this* then *that* will happen, if something is not done then a pre-planned corrective action can be taken – but in dealing with people cause and effect are less certain. Within certain very general parameters, a person's likely response can be predicted – if I, as the manager, encourage my staff members then they are likely to perform better than they would if I treat them with surly indifference. But there are many external influences that affect human response.

One of my colleagues has an assistant, Sally-Ann. A cheerful and generally unflappable woman in her mid-thirties, she is in charge of strategy for the company – a highly responsible position. She is very easy to get on with and seldom needs much help from her boss. Her team respect her enormously and enjoy a very good working relationship with her. However, one day Sally-Ann gave a team member a forceful reprimand for a very small error that in normal circumstances would have been ignored. Sally-Ann's uncharacteristic behaviour in this instance was the result of a string of unconnected external events.

Sally-Ann had had difficulty in starting her car that morning; she was not feeling well; it was raining and the heel of her shoe had snapped. So, apparently, had her temper. External events had produced a change in Sally-Ann's behaviour that resulted in an unpredictable and non-rational response – despite the need for her, as the leader, to demonstrate self-control and set an example to her team.

There is no way that a functional approach to people management would have been enough to restore the situation. Therefore my colleague had to take Sally-Ann aside for some careful counselling. Afterwards, Sally-Ann had to spend most of the rest of the day rebuilding her position with her team.

People are not machines and the many external factors that influence them mean that they cannot be managed and have to be led. We, as managers, need to remember this when considering whether we are managing in the right way. Leadership skills and competencies can be learned and a team leader's role or man-management position needs to have a profile set accordingly. Good leaders are valuable and, used correctly, will generate considerable added value for the company.

Have the right customer-focused products and services

This is where you need to consider two things: (i) what do you want to sell to the customers? And (ii) what do the customers want to buy?

The first question is answered by knowing what business you are really in and then working through the marketing strategy to determine what it is you want to sell. The second is answered by finding out what the customer really wants from you. With these two bits of information it should be possible to design your products and services so that they both meet the customer's requirements and bring in the profits you need.

Obvious? You'd think so, but in many companies the two items are not related. The company decides what it wants to sell and then tells the customer what they can buy – the 'Henry Ford approach to marketing' mentioned in Chapter One. The approach is not wrong, but it is full of dangers because it faces the customer with a *fait accompli*. It would be better if, having decided what it is you want to sell, you then package and present it to suit the customer's needs rather than your company's convenience.

About three years ago our old washing machine literally blew up and I had to buy a new one very quickly. My wife, who takes charge of such things domestically, questioned the salesman closely on what the machine would do – spin rates, wet load capacity, rinse cycles and the like floated in and out of the conversation – and she eventually chose one that would do what she wanted it to do. The machine was duly delivered and she settled down to make the thing work – an activity that involved much reference to the manual to decipher what symbols meant and dials did. We got there in the end, but it would have been much better if the switches had been clearly marked and the dials made more user-friendly. It seems the construction of the dials was determined by the clockwork control mechanism chosen by the company for ease of manufacture rather than by what the user would find easy to use.

Undoubtedly many people will say that their company *does* design products and services with the end user in mind. They will also tell you that they test the item with end users and modify accordingly. What they do not do, on the whole, is *involve the end user in the initial design and development phase* – yet, if they did, they would probably end up with a better design. Those companies which do involve the end user from the very beginning of the design process are generally making products that need to be highly customised rather than mass produced. Even in that field there is still room for improvement.

The external customer may be the only user of our products and services, but our staff also have to cope with our processes and procedures – especially those that relate to internal matters between departments. In these circumstances, our staff are internal customers and we should ensure that all internal matters are also designed with the end user in mind. 'But,' I hear you say, 'the purpose of an internal procedure is to gather data for the originating department or to allow control to be exerted.' Fine, but think it through: if the procedure is necessary (which is sometimes questionable) then accurate completion of forms, returns, and other data-transfer methods is vital. If the form has been designed for ease of completion, then it is more likely to be completed accurately – which will benefit everyone. On the other hand, if the form has been designed only for the ease of data input by the originating department, then the end user may well have difficulty in completing it accurately – especially if it contains the originator's jargon.

One such form was used within a large manufacturing company. Personnel wanted to assemble information on the number of hours worked in different payment categories so that they could calculate a person's wages quickly and accurately. The data they needed was the hours worked on a normal shift at the standard rate of pay, the number of hours overtime on that shift (at time and a third), the hours on additional overtime shifts (time and a half), on Sundays and public holidays (double time) and so on. To assemble this data they designed a form with many boxes, explanatory notes and formulas which, when complete, gave the exact amount to be paid to the worker. The form took, on average, about 35 minutes per week to fill in, which amounted to about 240 man-hours of time per week for the workforce of 400 – a loss of 30 man-days of productive labour. More time was wasted because of having to check data and search for missing information.

The form had to be redesigned and people from the shop floor were asked to help. The new form had five or six boxes that were filled in daily and which took, on average, one minute – a total time of five, maybe ten minutes a week. The income earning time saved was enormous – 170 hours or 21 days. The personnel department re-programmed the computer to accept this reduced data input and accurate results were produced just as quickly.

THE CONTINUOUS QUALITY IMPROVEMENT PROGRAMME

The CQI programme is based on the five areas discussed and is designed to ensure that continuous improvement is made possible in each of them. This is achieved through structuring and phasing the programme as in the schematic view on page 28.

As we work through and examine each part we must keep in mind the necessity of ensuring that

(a) the programme is accepted and owned by everyone in the company,

(b) each part contributes to the overall objectives,

(c) the programme delivers against its objectives,

(d) it is cost-effective and adds value, and

(e) it does not become an end in itself – it is, after all, a tool to be used to make your company a CQI business and thus ensure its survival.

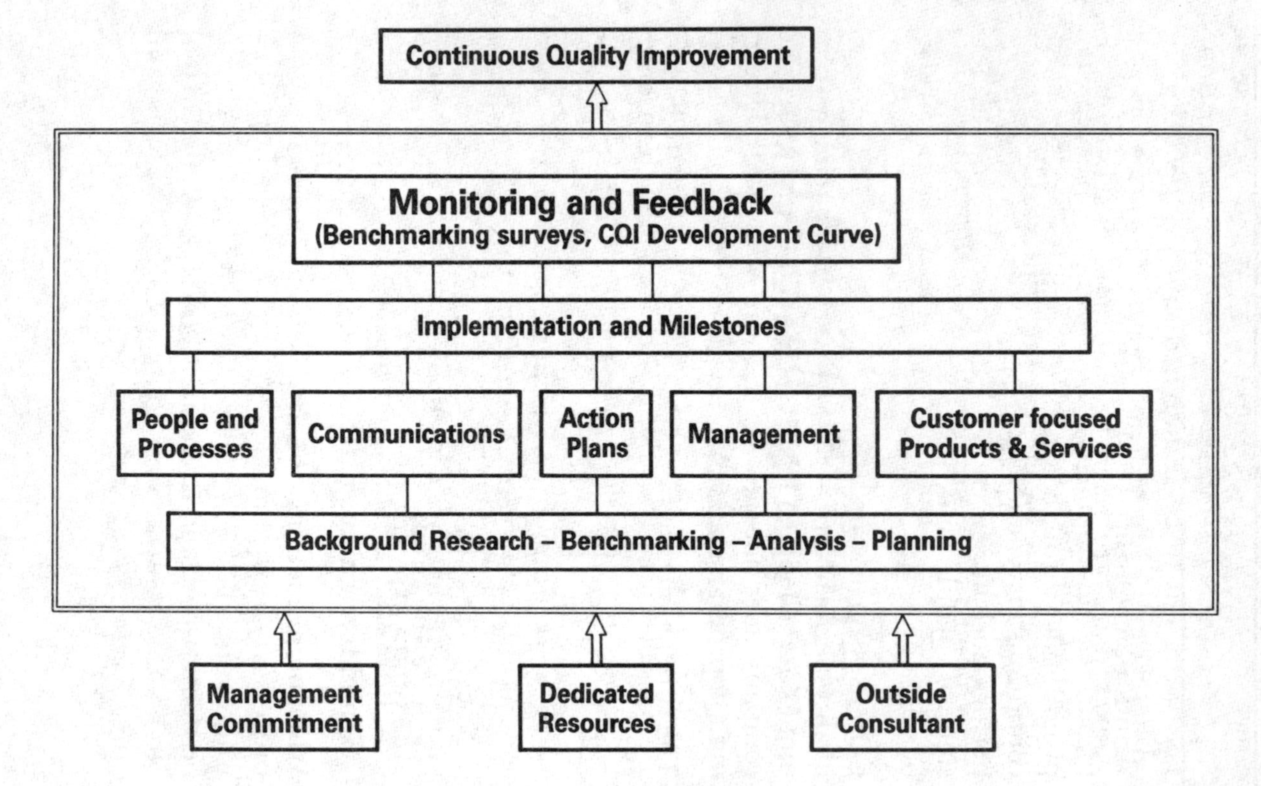

The Continuous Quality Improvement Model

SUMMARY POINTS

- 'Continuous Quality Improvement is a global approach to business development that establishes an integrated programme through which a company can achieve continuous incremental improvements in its chosen key performance measures by focusing on the better leadership of people and the improved management of business processes.'
- CQI uses quality management practices to obtain top quality performance – performance management techniques are allied to a process designed to make sure the company does not stand in the way of the staff delivering a top performance.
- The core of the programme is the business survival strategy:
    - determine the functions that need to be done
    - ensure the right processes are used in the right functions
    - get the right people to do the right jobs
    - manage in the right way
    - have the products, services, processes, and procedures that are customer-focused for both external and internal customers.
- When implementing the programme, remember that processes are rational with predictable outcomes, whereas people are so complex that their responses can be non-rational and cannot be accurately predicted.

Part II
First Steps

3

Establishing the Objectives

What do we want to achieve? • what do we want to look like? • setting the goals • where do we go from here?

ONE DAY you walk into your office and find that you are now heading the company's CQI programme. The decision to undertake the process has been taken and you must get on with it. At this stage the first thing you need to do is understand the reasons why the company wants to achieve continuous quality improvement. The reasons are obvious, you may think, but are they?

It would be nice to be able to believe that companies go the CQI route because they genuinely want to achieve improvements. Unfortunately, the motivation is seldom so straightforward. I have come across a variety of reasons why senior management pursues CQI. The most usual reason is that the company is perceived as slipping in its performance against key financial indicators and senior management want a quick and easy way of addressing this. On the other hand, they may also want to *appear* to be making the changes necessary for improved performance before being told to take action by the board of directors. In other words, a genuine reason for starting CQI is overlaid by a desire to look good in the eyes of the bosses. Sometimes this goes even further – the idea of CQI is raised by a senior manager who wants to score points with the managing director by seeming to be right up to date – after all *quality* is one of the buzz-words of current management thinking and everyone is doing it.

Occasionally, the reason for starting a quality programme is to obtain a marketing advantage – to obtain BSI 5750 or ISO 9000 registration is seen as a prerequisite for operating in certain markets. Once registration is achieved the whole cost is recorded as a marketing expense.

Understanding why it has been decided to undertake CQI will help you foresee what commitment will be made and what support offered by senior management. This perception is vital to the success of the project because if the backing is not there (because the reason for CQI is not genuine) then the programme will be allowed to die before it has really taken root. If you suspect that the reasons for CQI are not the right ones for success, then you must guard against lack of support and commitment and even, in extreme circumstances, refuse to start the process. At the very least, you will need to spend considerable time with senior management getting them to understand the implications of undertaking a CQI programme – after all, some of their jobs may be on the line as a result of rethinking processes and functions.

So, are there genuine reasons for CQI? The common reason – CQI as a corrective mechanism for improving performance against financial indicators – is actually perfectly acceptable because it links to the need for a company to continually reinvent itself and evolve if it is to survive. However, it would be even better if the company embraced CQI *before* its performance started to slip. A desire to reduce staff turnover or to improve performance from current staff is also a good reason, as is improving management and staff morale. Improving customer service (reducing customer complaints) and reducing time-to-market for new products are also valid and often quoted reasons. But the best reason of all is the simple desire to achieve continuous improvements in the way the company is run so that its survival is assured. If you are lucky, this will be the reason your company is undertaking CQI and if it is, then you can expect the support and commitment of senior management for what is a lengthy process but one with extremely satisfying results.

WHAT DO WE WANT TO ACHIEVE?

This question seems to floor many managers – it is as though the answer is too simple to be articulated. Yet if we are to set out on a journey then we at least need to know in which direction to travel, even if the final destination cannot be defined.

To make things easier we can re-phrase the question: **in what areas of the business do we want to achieve improvements?** Care must be taken that the glib answer – all areas – is not offered in place of thoughtful analysis of priorities. You can help the process of analysis by suggesting some key business dimensions so that thought is directed towards non-financial areas as well as financial indicators. I normally use a range of dimensions – or different sections within a given company – as this allows people to identify areas of achievement that build towards the overall objective of continuous improvement for company survival.

Human Resource Management

The first dimension I use is Human Resource Management but even this dimension is too big for most people to grasp, so I break it down into various component parts, all of which are related to the way we manage our people.

Overall staff satisfaction

The overall satisfaction level felt by the employees about the company as a whole is a good guide to the level of problems that exist within the business. Overall satisfaction can also correlate with levels of morale and, when combined with other data, can be used to check conclusions being reached in different dimensions. An improvement in overall staff satisfaction reflects improved morale, loyalty, job satisfaction, performance, and commitment amongst our people. It translates into improved financial results in that a happy company with high staff morale is usually a well managed and generally successful one.

Training

In this sub-category we are looking at how we are doing in terms of the formal training of skills and knowledge and in the coaching of those skills in the workplace. We can use the feedback from our people to determine whether training is sufficient and whether it is delivered at the right time. We can also look to see whether coaching is carried out to a satisfactory level in terms of regularity and frequency. Training is an **investment** in people and not an operating expense and, as with all investments, we need to ensure that we are receiving adequate returns. An improvement in training feeds through to overall satisfaction, improved performance, reduced costs and increased revenue.

Goals, feedback and recognition

Given the importance of these three subjects to performance management it seems only right that we should look at our use of them. An improvement in this sub-category will translate almost directly into increased profits and will affect financial indicators such as Revenue per Full Time Employee. To achieve an improvement, however, means that we have to improve the way we are managing our people and this has a much wider impact than a purely financial one.

Authority and responsibility

Here we are concerned with whether our people have the

authority necessary to do their job properly and whether they have accepted the responsibility of using that authority. An improvement here, normally by empowerment, comes through as better morale, increased performance, and reduced internal quality problems.

Career development

Companies that do well in career development terms generally have loyal and dedicated workforces and low staff turnover. These lead to reduced recruitment and training costs and the decline in performance caused by new people not being 'up to speed' is avoided. One of the top service companies in the transport sector has a staff turnover of 30% per year – it would be even more profitable if it had higher staff retention.

Job and working conditions

There are two distinct and separate items here, job conditions and physical working conditions. Job conditions are such things as the responsibilities and minimum performance standards and pay grades associated with a specific role, while working conditions cover the physical environment. Improvements in job conditions result in improved motivation and enhanced performance while improvements in working conditions result in greater productivity and higher staff satisfaction.

Service quality

This dimension covers how well we are doing now in terms of some of the basics of the quality movement. For example, external service quality covers things like customer satisfaction, problem-free service and product delivery, speed of response to customer enquiries concerning product usage. Internal service quality covers similar items within the company and focuses on interdepartmental relationships. Also included are management attitudes towards quality as perceived by the customers and by the staff, and whether service and service quality are part of the way we appraise our people's performance.

Customer focus

This dimension is really three interlinked categories relating to our customers and how we interact with them.

Sales focus

It is fashionable to state that a business is sales-orientated and is driven by the needs of the customer. This subcategory looks at the business as a sales organisation and allows us to see whether it is really sales-focused and customer-driven. Improvements in the sales focus of a business result directly in greater ability to meet customer requirements and increased sales.

The customer

In running a business we are always making assumptions about the way customers behave and what they want from us. We need to test those assumptions by looking at customer behaviour through the eyes of those who deal with them on a day-to-day basis and through the eyes of the customers themselves. Improvement in this subcategory allows us to make marketing judgements based on accurate customer information, thus increasing the likelihood that the customer will buy from us.

Marketing approach

Are we delivering the products and services our customers want in the way they want? Are we approaching the market in a way that will allow us to maximise the return on our investment? This is where we come to terms with what business we are currently in and what business we should be in. Getting the marketing approach right is a prerequisite for success and any improvement here will translate directly to the bottom line of the balance sheet.

Business development

While there is no right or wrong in terms of business development, we must decide what we want to be and how we want to interact with the customer. In this dimension we look at the development of our sales team, what drives the business (e.g., our technology, our processes, marketing, sales, or the customer), who drives the contact between the company and the customer, us or them, and what sort of customer they are – a commodity buyer, an interested prospect, or someone in need of customised solutions.

Internal relationships

This is another very big subject that can be broken down into certain key elements, all of which tell us about how well we are doing in managing the company.

Communications

Here we are interested in the internal communications between any individual, their manager and the manager above that. Communication is a two-way process which, in many companies, does not happen. By improving communications upwards and downwards we can ensure that everyone has the right information to do their job. This reduces the amount of time wasted searching for information and improves motivation and morale dramatically. (Think back to the UK's near-disastrous labour problems of the 1960s, 70s and 80s to see the effect of poor communications.)

Co-operation and teamwork

The way people work together to achieve the company's aim is of vital importance, if we can improve co-operation and teamwork, then less effort is required to achieve higher goals and increased productivity normally results. Looking at co-operation and teamwork also tells us a lot about the way we are managing the people and the business as a whole.

Capacity and resources

One of the great cries of the middle manager is that there are never enough resources and that capacity issues hinder the ability to grow. Is this a real issue or a reflection of poor management and incorrect utilisation of resources? Many companies have found that they really have *excess capacity* and better utilisation results in increased performance and greater output without an increase in overheads.

Support

This is a key area in which we look at whether the company is offering our people the support necessary to do their job, or whether lack of technology and management support gets in the way of our people doing their jobs correctly.

Key financial ratios

All companies have their favourite financial ratios on which they judge the business. Some that are popular include:

- Net Revenue per FTE in which the amount of revenue generated (after deductions of direct costs) is compared to the number of full time employees (FTE): a key indicator in a service company or people-intensive business.
- Revenue: Expense Ratio in which the gross revenue is compared to the total costs – if the answer is greater than 1 then the trading operation is profitable. This formula is related to Gross Margins.
- Free Cash Flow – will the company have enough cash flowing in to meet the costs of greater volumes.
- Sales: Fixed Assets – mainly used in capital-intensive industries.

- Dividends per Share – a measure of success as viewed from the shareholders' point of view.
- Change in Share Price – also a reflection of success from a shareholder's point of view although a poor indicator of a company's true worth.
- Breakeven Point – the point at which the company moves from loss into profit, which varies with volume of sales. (This is one area where mathematical modelling is vital.)
- Stock Value: Sales Volume Ratio, Value Added, Production Cost ratios – the list is almost infinite.

My advice is to work closely with your financial director to agree two or three simple but comprehensive financial ratios that offer a readily understandable overview of the company's true financial strength to all members of the staff.

If these are the areas of the business in which improvements should be made under a CQI programme, what about the improvement in our manufacturing, operating, administrative and other processes? Let me be quite clear. Improvement in the way we manage our people and approach the market will inevitably lead us to looking at our processes, but process improvement and re-engineering are an *outcome* of the CQI programme and not a primary focus. In deciding what it is we want to achieve, the areas in which we want continuous quality improvement, we are focusing on the people issues within the company – if we get these right then most of the other issues will resolve themselves.

What Do We Want To Look Like?

If we know the areas in which we want to achieve improvements, do we know what those improvements will look like? Do we know the objectives we wish to achieve? I have found that, in many companies, it is highly unlikely that anyone has really thought this through – mainly because very few people have a clear picture of what can be achieved – but it is something we must do before launching into the programme.

Much of the success or failure of any CQI programme depends on the communications strategy and the very first part of such a strategy is a clear statement of where we are going. Now, some managers try to define everything in terms of what has to be achieved by when and by whom. This is admirable and is a vital step that has to be taken **but not yet.** At the moment we are still at the stage of broad brushwork, not fine detail. What we need now is a vision of the future, a mission statement we can 'sell' to our people and behind which they can put their combined strength.

Your mission statement needs to be very precise but not all-encompassing, short and simple to understand, realistic but challenging, and it must tell your people where you are going but not how you are going to get there. I believe your mission statement should also tell everyone by when you plan to have achieved your objective. It is a statement of intent, a rallying point and a destination for the journey. As such, it has to be believable and provide a reason why people should commit to it.

Here are some examples of CQI mission statements:

- To achieve continuous improvement of 10% per annum in all key performance measures so that, in three years' time, we are the company against which our competitors benchmark themselves in terms of quality, performance, and management practices.

- To achieve continuous quality improvements so that, in two years' time, we are selected as the supplier-of-choice by all our tier one clients and 50% of all other clients on the basis of the quality of our service.
- To be recognised, in two years' time, as the company that potential employees most respect and would like to work for.

The first identifies that the company concerned wishes to be judged by its competitors, while the second chooses to be judged by the customer and the third by the staff. There is no reason, of course, why you should not combine all three to produce one statement. Once you have your mission statement you have the objectives of the CQI programme and you know what you want to look like by a certain date.

The next stage is to 'sell' the mission to senior management – which is not necessarily as easy as it sounds. Experience has shown that most senior managers want to have an input into the mission statement so that they can commit to it – this is a normal reaction based on goal setting: if the goal is set by those who have to achieve it then they will be committed, if the goal is imposed by others then it has little value and attracts little commitment. The problem is that too much input frequently results in a lengthy and often vague statement of intent that cannot be sold to your people. I recommend, therefore, that you consult all senior managers early on in the process and then undertake to produce the mission statement for their agreement. This approach recognises the need for input but relieves the senior managers of having to spend time working on the statement – with the result that your version is likely to achieve acceptance.

I cannot stress enough the importance of obtaining genuine commitment from the senior management team – they are all affected by the CQI process and it is their budgets that will have to find the money to fund the programme itself. To obtain this commitment, it is necessary to present the outcome of the programme in terms of benefits to them and their departments as well as to the company as a whole. Your aim should be to get everyone behind the programme: however, if this is not possible, it is vital that you get enough senior support to ensure that the programme is protected, the budget is likely to be approved and the CQI is not just this year's latest fad.

SETTING THE GOALS

By this stage you should have four things clear in your mind:

- why the company is embarking on a programme of continuous quality improvement,
- what the key areas for improvement should be,
- what the overall objective is, and
- the level of commitment amongst the senior management team to pursuing CQI.

The next thing to do is to start establishing the general goals of the programme.

Now, it might seem strange to be setting goals without a full understanding of where we are at the moment, but we are still a little way off launching the programme and the full proposal has yet to be completed. The proposal is the document that must be approved by the management team and must include an indication of the budget requirements. After all, no company embarks on a programme of major change without some understanding of what is involved and how much it will cost and it would be a pity to do too much detailed planning if the whole process is not to be given the 'go-ahead'. By looking at the general goals you will also begin to appreciate the benefits that will result as well as the obstacles in your way. So, what are the general goals?

Essentially, they are the outcomes of the programme and cover improvements in internal management practices, enhanced productivity and performance, better service quality from a customer perspective, improved cost/expense ratios, better internal relationships, more sales, greater market share, improved staff utilisation, increased capacity, better processes and so on. Should they be phrased in goal terms: what is to be achieved by when and by whom? Possibly not at this stage. It may be better not to quantify the goals until you have a full understanding of where the company is at present, but they can be stated as general objectives.

One outcome I always avoid is a discussion about potential cost savings through staff reduction. This is, of course, a possible outcome from the programme and one that many managers wish to see, but it is not what CQI is about. The fixation many managers have with cost savings is a negative attitude resulting from the short-term focus on financial results and it is almost always easier to find £500,000 of savings than it is to find £500,000 of increased revenue. A manager's job is not to reduce the size of the company, cut its capacity, and damage its ability to grow – it is to utilise resources correctly, manage for performance, and increase the business and the profits. A fixation with cost savings is the direct opposite of what a good manager should be displaying. 'But,' I hear some managers say, 'we have to get rid of excess costs to improve the bottom line.' What an admission of failure. It is time to break the mind-set of a whole generation of managers, get away from the narrow view of the balance sheet, and start managing properly. It is time to understand that **there is no such thing as excess costs, just badly utilised and badly managed resources** – our role as managers is to use the resources correctly and not waste them. No, I don't discuss staff reductions as a cost saving. I discuss the correct use of resources and the effect that can have on the balance sheet.

There is, of course, a perfectly valid argument for reduction in staffing levels, as discussed in Chapter Two – if correct utilisation of resources and modification of processes show that certain job functions are no longer necessary, then the people fulfilling those roles are no longer required. However, it would be better to reassign them to new jobs rather than remove them from the payroll.

WHERE DO WE GO FROM HERE?

With all this groundwork complete, and before developing the CQI Master Plan, we need to benchmark the business by finding out exactly where we are now. This is the subject of the next chapter.

SUMMARY POINTS

- Determine the real reasons why the company is to undertake a CQI programme – is it to achieve genuine improvements or is it to make senior managers look good to their bosses?
- Understanding the reasons allows you to judge the level of commitment there is to the CQI process.
- Identify the key areas for improvement – these will form the core of the CQI programme.
- Key dimensions include:
  - Human Resource Management
    - Overall Staff Satisfaction
    - Training
    - Goals, Feedback, and Recognition
    - Authority and Responsibility

    Career Development
    Job and Working Conditions
  - Service Quality
  - Customer Focus
    Sales Focus
    The Customer
    Marketing Approach
  - Business Development
  - Internal Communications
    Communications
    Co-operation and Teamwork
    Capacity and Resources
    Support
  - Key Financial Ratios.

- Process improvement and re-engineering are an outcome of a CQI programme, not the main focus.
- Establish a good Mission Statement for the CQI programme – one that clearly defines where you are going.
- Sell the Mission Statement to the senior management team to obtain commitment to the process.
- Set the general objectives of the CQI programme – these are the benefits that are the outcome of the process.
- Avoid, at this stage, any discussion of cost savings – especially those related to staff reductions. If cost savings and staff reductions come from the process, all well and good, but remember:
- **There is no such thing as excess costs, just badly utilised and badly managed resources – our role as managers is to use the resources correctly and not waste them.**

4

Where Are We Now?

Benchmarking • auditing against benchmarks – 'best practice' standards

YOU'VE WRITTEN your CQI Mission Statement, you've established the general objectives of the programme, and you've identified the key business dimensions in which you want to see continuous quality improvement. In fact, you now have quite a good idea of where you want to go. Having a destination means you can start planning on how you are going to get there and that requires another piece of information – **where are you now?**

You have to know where you are now if you are to have any idea of which way to go to reach the objective – even if that objective is only loosely defined. In today's fast-changing business environment it is difficult to be absolutely certain that where we think we want to be will ensure our success or even our survival. One CEO summed it up when he said, 'I don't know what we need to look like in five years' time – I only know it will not be the same as today.' But knowing what we look like today is important if we are to look different in five years' time.

ESTABLISHING OUR CURRENT POSITION BY BENCHMARKING

Knowing what we look like today means establishing a starting point from which we can work. To do this, we need to know

- what our existing customers feel about us,
- how we are seen by our potential customers,
- what our staff think about the company and the way we manage it, and
- what our key financial ratios show.

To find out this information we must do some research to 'benchmark' the business.

But first, a word of warning. You must not make any assumptions about your current position. Untested assumptions are not good foundations on which to build a successful programme. They are also dangerous in that there is a tendency in all of us to try and prove that our assumptions are correct and this may well lead to asking the wrong questions. Determining your current position must be approached with a totally open mind.

There are a variety of tools available to you – most are based on some sort of quantitative survey which can be very useful for 'benchmarking' the initial position and for monitoring performance. They also have a number of additional advantages:

- quantified (numerical) results are easily understood,
- they are accepted easily by most managers (especially those who tend towards a 'numbers' approach to management), and
- the surveys are repeatable thus providing directly comparable results.

A good deal of work has already been done in this field by a number of consultancies and other organisations and this can be used to provide outside comparisons and benchmarks to support and interpret your results. It also means you do not need to waste time re-inventing the wheel.

External surveys – what our customers and potential customers think

You need to consider two types of external customer – existing customers and potential customers. The first group can provide feedback on quality of service and on the products themselves while the second can tell you about how your company is seen in the marketplace. Both lots of feedback are useful and both will affect the CQI programme you are planning.

Our existing customers are those people who purchase the goods or services supplied by our company. In a manufacturing business they are the wholesaler, the retailer, and the end-user, which is a more complex chain than we would find in a retail operation where the external customer is the end-user. It is important that we fully understand who our external customers are.

My approach in defining the external customer is to sit with senior management and the marketing people and ask 'who buys our goods and services?' Next, 'who is the end-user of our goods and services?' This question is often more difficult to answer – any log jam that develops can often be broken by asking 'who might complain about our goods and services if there is something wrong with them?' Most managers find that the list of external customers is much longer than they first imagined.

Take a company which supplies sales skills training. When identifying their external customers they eventually decided on the following list for each course they provide: the salesman being trained, his sales manager, the Regional Sales Manager, the Sales and Marketing Director, the training manager, the Human Resources Director, the

Managing Director, and the shareholders. Excessive? No.

What the company was doing was identifying all the people in the client company who had an interest in the successful delivery of the training. That is not to say that all these people could actually answer questions about the impact of the training provided, and nor would they necessarily be asked to, but they are all affected by the product being sold by the training company.

Having identified the external customer, you must then decide what it is that you want to find out. The traditional service quality approach is to ask questions relating to satisfaction with the service or product provided, but CQI is about continuous change and continuous improvement and you should work with the marketing department to develop a comprehensive list of things on which you would like the customer's views. These may include design, colour, price, ease of use, cost effectiveness (i.e. value for money and cost of operation), service support, quality of delivery, packaging, instruction books – there are literally hundreds of items that can go on this list.

Going back to the training company, it identified the following items on which it wanted customer feedback:

- the way the course was sold (did the salesman find out what the customer really wanted, how well was it customised, the attitude of the salesman – did he try to pressurise, etc.);
- the way the course was taught (presentation skills of the seminar leader, technical competency, approach, the learning process, and so on);
- the course content (was it suitable, was it focused on the customer's needs, was it customised enough, was the course material satisfactory);
- finally, the added value provided by the course (increase in sales performance, increase in profits, enhanced motivation, improved image in the marketplace).

Clearly, the feedback being sought is much more than just 'are they satisfied'; in fact, the exercise is a piece of market research. It is designed to provide information that will help us:

- improve the way we interact with the customer,
- improve the quality of our product and the way it is delivered,
- understand whether or not our product is really customer-focused,
- identify training requirements, and
- identify weaknesses in the management of our customer interface people.

At this stage, we need not focus on what we are going to do with the information, just concentrate on acquiring useful data, and to do that, we need to determine the best way of getting the feedback. For example, you could try to get the customer to complete and return a questionnaire, you could carry out face-to-face and/or telephone interviews, or you could work in focus groups in which you call some customers together and ask them to discuss your service and products as a group. All these methods have advantages and disadvantages and you may well need to use a mixture.

Before selecting a method, or mixture of methods, consider how many customers you have and the number of responses you want. If, for example, your company has thousands or even millions of customers (and, remember, each sale may represent a number of customers) it will be impractical to carry out face-to-face interviews with all of them, but a sample of, say, 500 selected at random might be possible. Equally, a postal questionnaire survey may be possible across all the known customer base. But what happens if you are a retailer with thousands of unknown customers? You might consider a point-of-sale questionnaire or a random face-to-face interview immediately post-sale. Another consideration will also be whether you have the internal resources to carry out such surveys or will need to involve a market research organisation.

My advice is to consider using an outside market researcher for this initial survey of your existing customers unless you are dealing with small numbers (up to a thousand, for example). These organisations are specialists and, providing you have identified what data you want to collect, they can advise on the most appropriate methods. Remember, you need reliable data from a sufficient and representative number of customers so that valid conclusions can be drawn about the whole customer base. The statistical approach necessary and the methodology used is vital – hence the need to consider the use of specialists.

Whether you carry out your own research or decide to use an outside specialist, investing in a properly developed and tested research tool and a methodology that provides a statistically valid response will be repaid quickly, since you and your colleagues will be able to trust the information obtained and make critical decisions based on it.

Whichever approach you have decided on, the first thing you need to do is identify the key elements about which you want information. Unless this is done, you are likely to end up with so much data that understanding what it tells you becomes very difficult.

Let's use the training company discussed earlier and see how a strategy could be developed.

Although it identified a wide range of end users, the company really needed feedback on four things:

- the way the course was sold
- the way the course was taught
- the content of the course
- the added-value provided by the course.

The way the course was sold

Here they needed to know whether the account executive had identified the customer's needs correctly, whether they had proposed a suitable solution that satisfied those needs and whether the benefits of the solution were presented in a clear and understandable manner. Finally, did the customer consider the account executive had behaved in a professional manner and represented the company correctly? Clearly, these questions could only be answered by the person who had booked the course – normally the training manager or human resource manager.

The way the course was taught

The company wanted to know whether the teaching approach was correct for the material and the delegates, whether the seminar leader had been professional and clear in his presentation, whether he or she was credible and acceptable to the delegates, and whether they were satisfied with the way the content had been presented. These questions could only be answered by the delegates themselves.

The course content

The strategy here was to ask whether the content was addressing the needs of the delegates as far as the company was concerned and also as far as the delegates were concerned – not necessarily the same thing. Feedback was also needed on whether the content was pitched at the right level, whether it was clear and understandable, and whether it was correctly supported by the course material. These questions were best answered by the line managers whose staff attend the course.

The added value of the course

Here they needed to ask about the pricing of the course and whether the customer felt it delivered value. They were also interested in the scale of the investment return (i.e., did the training produce increases in productivity and revenue). These questions had to be answered by the senior management of the customer. The company was also interested in whether the delegates felt they had benefited from the course.

Given the chosen strategy, at least two different groups of respondents had to be surveyed – the training manager and the delegates – and this required two different approaches. With the training manager the telephone was used, with the interviewer completing the survey by asking questions and noting the answers. The delegates were given a questionnaire at the end of the course which they were asked to complete and return to the seminar leader before departure.

As a one-off activity, this type of survey has little value, but by repeating it for every course run over a six-month period the accumulated data showed trends from which benchmarks could be established and areas for improvement identified.

Obtaining feedback from potential customers requires a similar approach, but here the research produces information that is primarily of use to the marketing department. Even though of no direct importance to the CQI programme as a whole, market research is valuable and necessary: I have found that, as the CQI programme takes effect, there is a need to develop new approaches to the market and, for this, knowing how the company is perceived by potential customers and what products and services are required by them is vital information.

Whether you should carry out your own market research or employ an outside consultancy is a matter of choice, but most companies find that using outside specialists more than repays the investment. However, market research companies tend to have their own individual approach and, frequently, a set method of carrying out the work – so shop around. You need to make sure that the questionnaire is based on what you really want to find out, that the research company will test it thoroughly, and that the results will be presented in a way that is easily understandable and lends itself to being used without further analysis.

This last point is important for *you* – being able to present the results of the research in their original format is a powerful tool in getting the information across to your colleagues because people tend to place more credibility on results produced by independent consultants than they do on those presented by a someone in-house, especially if bad news is involved. Finally, make sure that the database will be retained in case you want more detailed analysis and that it will not be used by the research company for any other activity without your express permission.

One of the great advantages of market research is also one of its biggest hazards – the quantity and range of analysis produced can be awesome. But is it necessary? Many research companies seem to believe that more is better but I have found that more tends to obscure the issues and people will spend hours debating small tactical information rather than focusing on the strategic issues and the key benchmarks.

Internal surveys – what our staff think and what the financial ratios show

In Chapter Three I outlined the key dimensions involved in internal benchmarking – human resource management, service quality, customer focus, business development, internal relationships, and key financial ratios, all of which have sub-categories. All these, with the exception of the financial ratios, are 'soft dimensions' in that they deal with behaviours within the company, and the best way of testing them and establishing 'soft benchmarks' is to carry out a staff survey.

Like all specialist tools, staff surveys require care in their application and skill in their use. Inappropriate construction of the questionnaire and inaccurate application will lead to frustration amongst the work force, resulting in a poor response and a waste of your budget resources. Inappropriate or badly constructed analysis will also reduce the benefit that can be obtained. In the circumstances, it is vital that you ask yourself the following questions:

Do I survey all the staff or just a statistical sample? Clearly, the best way to obtain a true picture is to survey everyone, but a statistical sample may be more appropriate if your company employs a very large number of people. There are, of course, advantages and disadvantages to each – if you go for a full survey you could be faced with many thousands of responses that take time to process; on the other hand, a statistical survey is harder to set up if you are to ensure that the sample is truly representative (a point I will discuss in a moment) and may be expensive and difficult to administer.

One of my clients employed 10,000 people and his initial idea was to do a statistical survey of 20% or 2000 respondents. Before we went ahead on this basis, we considered the likely response rate – the number of questionnaires that would be completed correctly and returned. Past experience shows this to range from 30 to 80% depending on the staff climate within the company. In the present case, previous surveys had attracted a 40% response rate, which would give us 800 responses – less than 10% of the total staff. I felt this would not be a representative sample and we would not be able to base future business decisions on the results.

A further complication would arise. The staff were grouped in 10 principal functions across 13 divisions: we would have had to take 20% of each function in each division, which would take some doing. With the likely response rate, it was very possible that all the sales staff would answer the questionnaire and all the backroom boys would not, leaving the result of the survey heavily skewed.

The only time I think a sample survey is a reasonable possibility is if the survey population is all in the same function and the same business division. In all other circumstances I recommend a full survey. (In this particular example, we surveyed all 10,000 and achieved a 76% response – giving a very robust set of data on which serious business decisions could then be based.)

Should the survey be totally anonymous or should the respondent be identified on the form?

If you want a worthwhile response rate and honest answers – and thus robust data – the answer is to make the survey totally anonymous. Asking people to identify themselves provides you with nothing in terms of useful data and is likely to depress the response rate to unacceptable levels.

Should the survey be run internally or by an outside consultant?

This is linked to the anonymity question – a survey run internally will always be regarded as suspect as far as anonymity is concerned. Frankly, it would be foolish to conduct a staff survey yourself – you should always use a specialist consultancy. This does have other advantages: most specialist consultancies have access to the scores from other surveys and can offer comparative results that can be used for benchmarking. They also have the experience necessary to carry out the analysis and to interpret the results, and they can present the results impartially to your senior management team, leaving you free to drive the CQI programme and action planning.

Should the survey be carried out by everyone at the same time?

This is a very good question. In practical terms you may need to give everyone a week or two to carry out the survey (so that they can all be contacted and given a form), but I certainly recommend doing the complete survey all at the same time as this can increase the response rate. Naturally, the questionnaire will be completed in company time, since it is a company activity.

How long should the questionnaire be?
This depends on what you want to find out. Survey questionnaires of 100–130 questions and taking about 30–40 minutes to complete are quite normal.

What sort of questions should be asked?
My preference is for statements with which the respondent is asked to agree or disagree on a sliding scale. For example:

When my manager gives me feedback it is done in an appropriate manner

5 4 3 2 1 0

The numerical scale is 5 = strongly agree, 4 = agree, 3 = neither agree nor disagree, 2 = disagree, 1 = strongly disagree, and 0 = Not Applicable.

You should also include 'reverse questions' or 'negative questions'. These are statements whereby the desired answer is a negative (i.e., disagree or strongly disagree). For example:

As a business, we do not consider our customers' needs enough

5 4 3 2 1 0

'Reverse questions' are there to ensure that the respondent thinks about the question before answering and does not just answer all the questions the same way. If 'reverse questions' are to be included then they should represent in the region of 20–30% of the total questions.

Should open-ended questions be included?
Some people place great faith in open-ended questions, believing that by asking people to write down their opinions they will obtain useful or honest information. In practice, this is less sure – many respondents will not

complete open-ended questions because they believe that their anonymity will be threatened. You may say that, in your company, everyone gets on well and worries about anonymity are misplaced – you may be right, but I doubt it.

Experience has shown that open-ended questions do not significantly affect the response rate but they do tend to be skipped. If they are answered then some useful information is obtained which can be analysed by themes, but the extra information obtained seldom seems to justify the extra work involved in correlating the answers. Open-ended questions should not be ruled out, but the pros and cons must be weighed carefully.

How should the questionnaire be organised?

Here I am talking about the internal arrangement of the questionnaire. The questions should, of course, be grouped so that they are logically arranged as far as the respondent is concerned. This will need checking thoroughly and you should test the questionnaire on a small but representative group of staff. Be prepared to alter the order if it will encourage a greater response.

At this stage, you should discuss with the consultancy who will administer the survey how the results will be analysed. The questions in the survey will be arranged in a way that is easy for the respondent to complete, but the analysis will have to be in the 'soft' dimensions you have selected. This will require you and the consultant to agree which questions refer to which dimension – some questions may well fall into more than one. For example, the question: 'How satisfied are you with your involvement with the decisions that affect your work?' would fall into the following dimensions: *Overall Satisfaction; Goals, Feedback and Recognition; Authority and Responsibility;* and *Internal Relationships – Communication.*

You should also agree the method used for presenting the results. The easiest to understand and to use is a bar graph shaded to show the percentage of answers that scored positive, neutral or negative. The example below is taken from a recent survey I ran for a client.

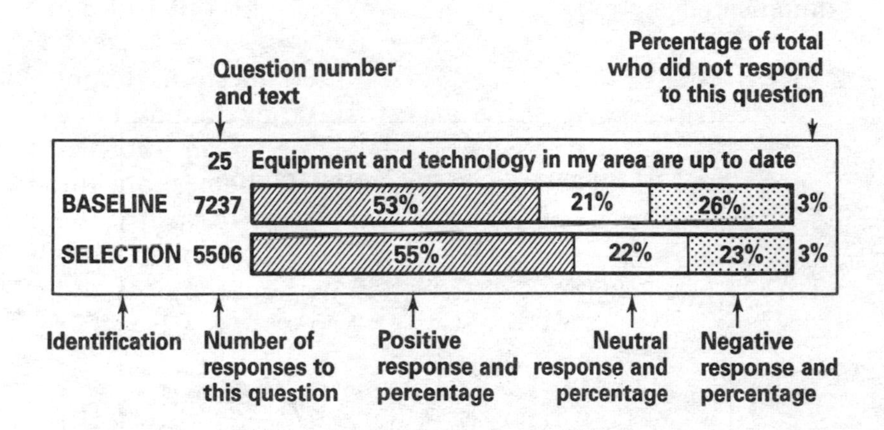

One way of presenting staff survey results. The Baseline graph shows the scores for all the staff whereas the Selection graph is for a specific group within the survey population.

As far as presentation of the dimension scores is concerned, I favour the use of a pie chart as illustrated below. The benefit here is that it is easy to see what the scores are and by 'exploding' the negative score you can highlight the problem area. I find that senior management prefer this approach, rather than receiving vast quantities of numbers that mean little to anyone except you and the survey analysts.

The dimension scores, which are made up of selected questions, are designed to summarise the scoring and provide a benchmark against which you can soft benchmark the business. The dimensions chosen, as discussed previously, are those which are the overall focus of the CQI programme and they are the ones in which the changes are measured. The initial scores established in the research phase act as the starting benchmark and at periodic intervals during the programme (probably every 18 months) you will survey the staff again to establish how well you are doing. To do this, you would use exactly the same questionnaire with the questions allocated to exactly the same dimensions.

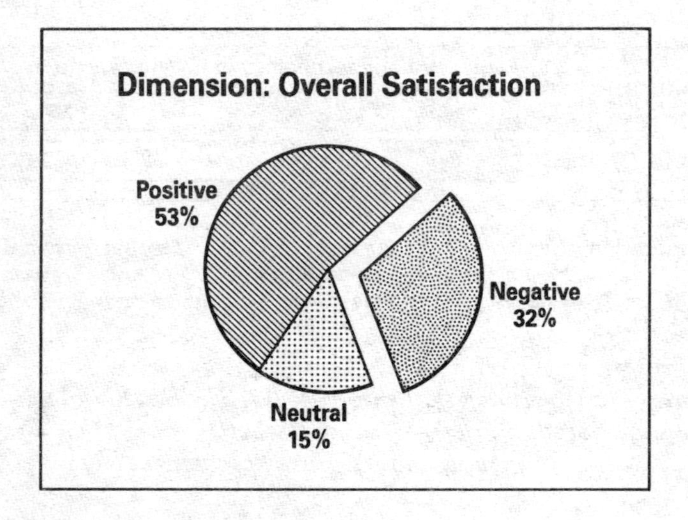

Presentation of a dimension score for use in benchmarking

I am frequently asked why the survey is repeated at 18-month intervals, rather than, say, 12 months or 24 months. In general terms, it is a matter of practicality. Once the questionnaires have been returned to the consultant the data has to be loaded into the computer and a database built. Depending on the size of the survey, this can take anything from a few days to a few weeks. The process of analysis can then take a month or so to reach a stage where it can be presented to the client's management.

Following presentation, it is then necessary to 'cascade' the results to all members of staff – this is vital to the communication process within the company, it helps build trust, and it sets the scene for the CQI programme and action plans. Following 'cascade', action plans have to be prepared, implemented, and allowed to produce a result. Overall, the process from survey to full implementation of action plans takes around 6 months and then a further 12 months of effort should be allowed before surveying again to check progress.

But why not survey every two years and really allow the action plans to work? Well, we all need to know on a regular and frequent basis how we are doing against our goals and waiting two years is just too long, especially since the results of the second survey will not reach everybody for a further six months. We also need to take corrective action if progress is not according to plan. Surveying every 18 months is a compromise between the need for constant performance feedback and the practicalities of providing that feedback.

Many people assume that the financial side of the company is purely the concern of senior managers and the finance director. How wrong they are. The financial health of the company is everyone's concern and monitoring that health is vital. It is also vital for the CQI programme – the outcome of a successful CQI programme is an improvement in the financial health of the company and monitoring the key financial ratios is part of the way we monitor the progress of CQI.

But there is a problem. Financial information tends to come in indigestible formats – especially for the financially uninitiated. This means that much of the financial data available will be unintelligible to your staff – hence the need for a few key financial ratios. As suggested in the previous chapter, you should work with the Financial Director to establish around five Key Financial Ratios (KFR) which will be used to judge the health of the company and to measure the return on the CQI programme. Naturally, once you have your KFRs you will need to benchmark them to establish the starting position.

AUDITING AGAINST BENCHMARKS

Since any CQI programme is about improvement within a single organisation, auditing external benchmarks is helpful but not vital in terms of internal behaviours. Of course, it is interesting to know whether a 35% negative score to the question '*The amount of work I'm expected to do makes it difficult for me to do my job well*' is good or bad (actually, it's the average score for many businesses) but the real issue is whether you can improve that score and reduce the negative to 30%.

On the other hand, there are certain benchmarks that establish whether a company has reached a certain standard in quality terms. Such benchmarks can be found within the ISO 9000 or BSI 5750 programmes, within 'best practice' standards, and from the general requirements of Performance Management. Although I have great respect for the ISO/BSI concepts, I do not believe they should be regarded as the be-all-and-end-all of good practice – unless you are looking to register under ISO or BSI as a quality company: it is much more important for you to decide what will constitute good practice within your company.

Generally speaking, you should identify a number of key standards of the 'best practice' variety which are applicable to your company and your CQI programme and you should benchmark your company against them. For example, in the Human Resource Management field most quality programmes require that every person has a written job description. In my book *Managing for Performance,* I go further – under Performance Management, not only should everyone have a written job description but Minimum Performance Standards (MPS) should also be included. This document should be agreed with the job holder as the basis of the job and the MPS as the basis for continuing to hold the job. The goals you apply to the job to obtain greater performance are not part of the MPS and do not form part of the job description. Having selected

Job Descriptions as a 'best practice' standard for your company you can now check to see if the company is meeting the standard.

Other standards might include:

- a process for performance reviews
- a process for annual appraisals
- a training programme associated with each job function (from the lowest worker to the highest manager)
- a coaching programme in all functions and for all levels
- measurable or observable goals for all job functions and job holders
- a clear career structure for each job function
- a clear career structure for each person
- a clear and understandable pay structure known to everyone
- an external service quality process (i.e., regular customer satisfaction surveys, quality circles, quality reviews)
- an internal service quality process
- a customer focus review process (are we selling the right things, in the right way, to the right people, and at the right price)
- an internal communications strategy
- a capacity review process
- a financial review process (including key financial ratios known or available to everyone).

This is not an exclusive list and you may well have others that you want to include. The key to selection of standards is to identify those that reflect the way you would like to be. The ones above are just a sample of those that every company should have – the final choice is yours.

At this stage, it is sufficient to identify the CQI standards you will apply and to check to see if the company has already achieved them. But what happens if you have, say, a process for annual appraisals but it is not a very good one or is badly applied? My suggestion is to add this to the items for review and that need action plans – only when you have a good process and it is properly applied can you say that you have achieved the standard. Having said that, I also believe that you should not try to fix those things that are working well at the moment. If, in our example, the annual appraisal process is reasonable and is generally applied then accept this as a starting point – a benchmark against which further improvements should be sought.

It is all too easy to get bogged down with standards to be achieved and things that need doing and the problem then becomes one of prioritisation. You should certainly identify the standards you want to achieve and you must benchmark the company against them, but at this stage that is all you do. Writing better processes, establishing better standards, is all part of CQI – **once you know where you are now.**

SUMMARY POINTS

- To determine the way ahead you need to know where you are now.
- Establishing the current position means knowing:
  - what your existing customers feel about your company
  - how your company is seen by its potential customers

- ▸ what your staff think about the company and the way it is managed
- ▸ what the key financial ratios show.

- Use quantitative surveys for benchmarking – they have a number of advantages:
  - ▸ quantified results are easily understood
  - ▸ they are accepted more easily by most managers
  - ▸ they are repeatable, thus providing comparable results.

- In planning an external survey of customers first define exactly who the customers really are – they are the ones who obtain a benefit from your goods or services.

- Identify exactly what you want to find out – a properly constructed survey helps you understand:
  - ▸ the way you interact with the customer
  - ▸ the quality of your product or service and the way it is delivered
  - ▸ whether or not your product or service is really customer-focused
  - ▸ the training requirements of your people in terms of delivering the product
  - ▸ weaknesses in the management of your customer interface staff.

- Choose carefully the methodology of the survey and seek advice from experts.

- Use outside experts for the survey of potential customers – market research can be complicated and market research agencies know what they are doing.

- Before surveying your staff, determine the dimensions you want to benchmark.

- Ensure that your sample, multiplied by the likely response rate, will provide a valid response that is representative of the whole company in terms of both size and mix. Better still, survey everyone.

- The best response rates are obtained if the survey:
  - has been tested for ease of completion
  - is run by a neutral outside consultancy
  - is completely anoymous
  - is completed in company time.
- Survey questionnaires can be quite long – 100 to 130 questions is quite normal and, if answered by a scale of preferences (strongly agree – strongly disagree), take only about 30–40 minutes to complete.
- The survey should contain around 20–30 'reverse' questions in which the desired response is negative (disagree) – this acts as a control.
- Open-ended questions – ones in which the respondent is asked to write an answer – are difficult to analyse and do not necessarily provide useful information.
- The survey results should be presented graphically as this is the easiest to understand. Presenting dimension scores against which you are benchmarking (rather than just the scores for individual questions) helps in understanding the results.
- The survey should be run every 18 months – this provides a compromise between the need to obtain feedback and the practicalities of providing the feedback given the need to allow time for action plans to have an effect.
- Ensure that Key Financial Ratios are established to monitor the financial health of the company and the financial return from the CQI programme.
- Select a range of 'best practice' standards that suit your company and include these in the overall benchmarking process.

Part III
The CQI Programme

5

Developing the Master Plan

Selecting the CQI team – leadership – team roles – team building • the CQI Master Plan – first steps – milestones – development curve

WITH THE RESEARCH PHASE successfully launched, it is time to turn to developing the CQI Master Plan based on your findings. This part of the process is still task-based and it falls to you, as the person who is leading the project, to draw up the strategic plan and to set out the short-term objectives and longer-term goals. It is also your responsibility as the leader to draw up a list of resources you will need – people, budget, support – and to obtain the commitment of senior management to ensure the success of the programme.

The first step is to establish a small strategic planning team (perhaps seven or eight people) and establish who is to lead it on a day-to-day basis. This does not necessarily mean that you will give up the leadership – you will remain responsible for the programme and its objectives – but you need help and the actual planning process may be better led by someone else. After all, you will need a team of people, some working full time on CQI and others assisting on an *ad hoc* basis and you are likely to have other responsibilities.

SELECTING THE CQI TEAM – LEADERSHIP

If the programme is to be successful it needs to have an overall leader, a person who owns the objective and can and does take overall responsibility. This may seem rather obvious, but a surprisingly large number of programmes are launched without a leader in place.

So how do we select the leader?

First, let me explain what the leader has to do. I believe that the best approach to leadership is to consider it as a function – a job that has to be done – and the best explanation of this was put forward by Professor John Adair. Calling his approach 'Action-Centred Leadership', Adair concluded that it is what the leader does rather than what he or she is that defines leadership. He went on to suggest that all leadership actions would fit within one of three areas:

- Achieving the *task*
- Building the *team*
- Developing the *individual*.

If the balance between these three critical areas of leadership action was maintained then the leader would be effective; if the balance was not there, then problems would arise. Adair represented this theory as three overlapping circles as illustrated on page 73.

When managers attending my Performance Management Workshop are asked to complete a profile based on Adair's work (see my book, *Managing for Performance,* Chapter 4) many of them show heavily skewed results with most of the focus on *team* and *individual* and with very small *task* focus circles. During one-to-one discussions with these managers I have found that the lack of *task* focus is almost always a result of the managers having been told that they

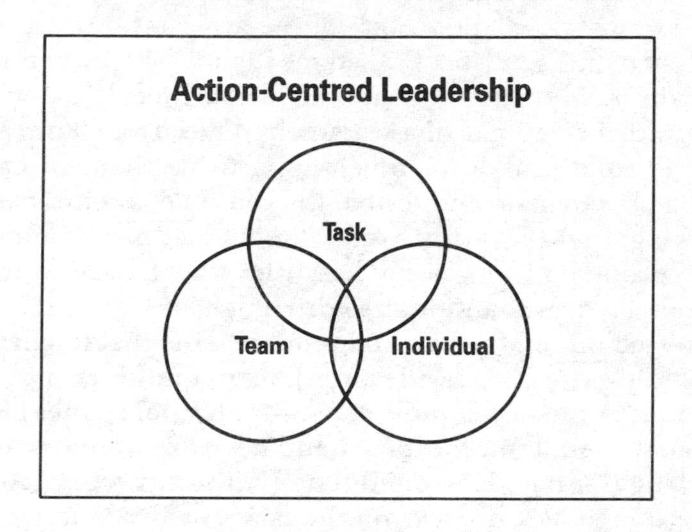

Adair's Action-Centred Leadership diagram showing the three overlapping circles in balance.

must be team players and that people are the key. This is, of course, absolutely right. But a leader cannot afford to indulge in people management at the expense of achieving the *task* he has been set. A leader's prime responsibility (if one can be defined) is to achieve a given *task* or objective – to do this, of course, he will need to build a *team* and develop the *individuals* in the team until the team is effective and can achieve the *task*.

One young manager described what had happened to him: 'My job was to lead a product development team and we had been tasked with developing a new piece of technology to enhance the effectiveness of one of our machines. All the way through my training the management theory was to focus on team building, on having a happy team that worked well together. No one had ever told me to think first about the task.

'Based on my experience and the prevailing thinking, I pulled together a team of experts and we set to work – we investigated all sorts of new ideas, we tried a range of new linkages, we developed state-of-the-art concepts. After about six months, my boss called me in and asked when we would be ready since the upgraded machine had to be tested with a customer in a few weeks' time. I was stunned – we had months of development work to do, then a prototype had to be made and tested, we would be lucky to have a working upgrade ready inside a year.

'I explained all this to my boss who was, I have to say, considerably less than impressed. He left me in no doubt that my job was on the line and reminded me that the original objective was to provide a working upgrade and not to have a nice time developing state-of-the-art solutions. This time he spelled it out for me – I and my team were to have a working upgrade ready and tested within six weeks. Any failure due to lack of focus on the task would result in me being given opportunities to pursue my career with another organisation.

'I got the message. The team got the message and, being nice guys, they dug me out of the hole we were in and we produced a first-class working upgrade inside the time frame although it meant working 18-hour days and seven days a week for six weeks.'

This young manager had failed to take full responsibility for the task, despite being the leader. He had not defined the team's objective clearly – including when it had to be done by – and he had not kept his team focused on achieving it. Now, you might say that his boss was also to blame for not having clearly defined the task in the first place and it is certainly true that some fault does exist there. BUT, the first responsibility of a leader is to take ownership of the task, define it clearly for the team, understand it completely, and never, never lose sight of it.

This is what you must do with the CQI programme. As the person designated to run the company's CQI programme, it is your responsibility to act as the leader. And the first thing is to take ownership of the task and the objectives. In general terms, this has already been done.

You have an objective to achieve continuous quality improvements (therefore there is no final time frame) and you have defined the principal areas in which improvement should take place. So, at least you know where you are going and you can and must remain focused on this task.

But achieving a task requires careful planning and it needs people – so which comes first? My suggestion is to break the task down into understandable chunks. The first chunk is to prepare the strategic plan and to do this you need planners.

SELECTING THE CQI TEAM – THE STEERING GROUP

Most managers would now look around to see who was free to work on the CQI project. Some may even look to see if their friends are available to help. A *leader* would first decide what skills and competencies are required within the team for it to be successful and then go out and 'recruit' the necessary people from amongst the existing work force. At this stage you must also look ahead and, remembering that programmes fail if they are imposed and not owned by the staff, you must include in your requirements not just functional skills but also the right personalities and a wide and varied range of people selected mainly from the workforce rather than the management levels.

Given the immediate objective is a strategic plan, the functional skills you need in the team will include:

- logical and analytical thinking
- attention to detail
- conceptual or strategic thinking
- clear presentation abilities.

In addition, you should look for certain competencies and attitudes:

- willingness to think about the unthinkable
- willingness to challenge the accepted status quo
- the ability to question beyond the normal
- the ability to understand that simple concepts may produce better solutions than complex ones
- the breadth of vision to ask 'why?' – why do we do this job? why are we really in business? why do we need this function? – and not to accept the simplistic or obvious answers.

Combined with the functional skills and competencies you need to consider the personalities of those involved. Do you want a team made up of individuals all working on their own parts of the programme or do you want people who work best as a team and who address each issue as a group? Do you want goal-orientated people or those who prefer to work towards a team goal? And what about people who will act as devil's advocates and challenge the team? Only you can decide what sort of a team you want to have as you develop the CQI Master Plan.

When selecting team members we all tend to look first among those people who are like us – maybe they have the same pay grade, the same corporate rank, come from the same sort of background, or perhaps think like us. You must resist this tendency and consider a wider group from which to select your team. Consider bringing in customer interface staff, back office workers, storemen, secretaries, marketing people, junior and senior staff, new recruits and long-service employees, middle managers and senior managers. Don't forget to consider men as well as women and any other mix of groups such as minorities (different races, the physically disabled, etc.). The more the team members reflect the mix within your company the more likely it is that 'ownership' of the project will be taken by those involved.

Finally, how many people do you need on the team? Since the objective is a strategic plan you could have just one or two people – equally, you may go for a bigger team knowing that eventually the strategic plan will have to be broken down into action plans that have to be implemented.

Some years ago I made the mistake of starting out on a CQI programme with a very small team of just three people. We all worked well together and the development of the overall strategic plan was easily accomplished, but the problems started when we had to begin the process of breaking the master plan down into tactical action plans. We found that we were soon overburdened and lacked the range of skills necessary for this second stage activity. Naturally, we brought in new members as they were needed, but the team never really worked well again. The new people lacked the knowledge obtained during the strategic planning phase and had missed out on the background discussions that had gone on. The original team had to spend a great deal of time covering old ground to bring the others up to speed, so no one was getting on with the work that needed doing and the original team members resented the newcomers.

I should have assembled a full team right at the start and assigned tasks as required. In this way everyone would have been involved in all the background discussions and all the work would have been covered.

I suggest that you should pick a team of around seven to nine people with as wide a mix as possible. Included in the team should be the functional skills required as well as the range of experience reflecting the main functional departments (such as HR, sales, marketing, production, administration, etc.). By choice, I would include one senior manager (management team or main board level) and one middle manager and bring in the others from the rest of the workforce, not forgetting the non-line people such as secretaries and administrative personnel.

Selecting the CQI Team – Team Roles

The team will only work successfully together if everyone has a clear understanding of what each person should be doing. This means that you will initially have to allocate roles to your team members and help them develop in those roles.

At this level, you must take the role of team leader. You own the task and you understand the objective, so it is your responsibility to develop your team to fulfil the task. As the work progresses, you may choose to appoint or select a different leader for a specific task, but you will remain the overall leader of the CQI team.

Let's look at some of the other team roles that need to be undertaken.

The Godfather

In any programme that is likely to have a profound effect on a company, it is vital that it has a 'protector' – a person of sufficient rank to ensure that agreed budgets and actions are protected from cuts and from re-prioritisation. Programmes that change the way things are done in a company attract suspicion and fear from all levels, including management. Most of the programmes from the 1980s resulted in people losing their jobs or having them radically redefined – down-sizing, right-sizing, de-layering have all left a residue of fear, as have TQM and business process re-engineering. In my experience CQI tends to attract the same fears amongst management in particular.

When managers feel under threat, it is natural for them to try to protect themselves from the effect of the programme and they will either consciously or unconsciously resist what is being done. Some resistance is likely to be in the form of lack of co-operation, reluctance to release staff for participation, failure to implement agreed changes in staffing levels or functional responsibilities, failure to supply information when requested, and an attempt to restrict the influence and scope of the programme. While many of these tactics can be overcome through a good presentation of the benefits of what is proposed and a clear communication strategy, it is false hope to think that everyone will welcome changes that are good for the company if they are not as clearly good for the individuals concerned. This is where the 'godfather' comes in.

The role can only be carried out by someone at the very highest level, someone who can intercede on behalf of the CQI team with senior management and can give clear instructions to the managers who are resisting. If you have not actually recruited such a senior manager to the CQI team then you need to find someone on the outside to whom the team can report.

The 'godfather' also acts as the interface between the CQI programme and the senior management. He or she can protect your budget, get you extra resources, obtain authorisation and commitment to implement changes and actions. On your part, you will need to keep your 'godfather' well briefed so that he can carry out his responsibility to the programme.

In 1989 and 1990 I was putting in a CQI-type programme for an international bank and I met a great deal of hostility and resistance. Fortunately, my 'godfather' was the business manager (the head of the business division) and he protected me and my team and allowed us to do what was necessary. Some years later I found out just how much 'godfathering' he had done – several of the senior management team were so fearful of the effect of the programme on their departments that they tried on a number of occasions to have the whole thing cancelled. The business manager protected the programme and the outcome was a dramatic increase in the productivity (and thus the profitablility) of the business – though a number of senior managers found themselves pursuing their careers elsewhere in the bank.

The Rapporteur

This role can be filled by anyone, but I recommend you appoint someone with good presentational skills because it is the 'rapporteur' who is responsible for maintaining the records of the programme and for presenting the results. Some people assume that the presentation of the results is the responsibility of the leader (you), but it is frequently more effective if a junior person is used in this role – it helps develop them as an individual, it enhances ownership of the programme, and it demonstrates and illustrates the commitment of all levels.

My choice has always been to use someone who does not normally come in contact with senior management or have to make presentations to large groups. In one case we used a machinist from the shop floor and on another occasion we used a junior pay clerk from personnel. In each case the effect was good – the individuals developed as people, commitment was high and, providing I or one of the others was around to answer the policy questions, it was easier for other low grade employees to accept what was being said when it came from one of their own.

The 'rapporteur' should be supported by someone with good writing and graphics skills to ensure that all presentations are of the highest quality and are consistent in style and layout. I return to this topic in Chapter Six.

The Devil's Advocate

Not an easy role by any stretch of the imagination and many people find it very difficult to play it dispassionately. The main purpose of the 'devil's advocate' is to try to find anything and everything that is wrong with a proposal. They do this by questioning and arguing with the proposers who have to defend what they are doing. If they are unable to convince the 'devil's advocate' of their position, then the proposal is flawed and should not be adopted.

Given the adversarial nature of the 'devil's advocate' you have to pick this person with care and you may need to have everyone play the role in turn. It is also a role in which people must be coached so as to avoid the introduction of hostility or personal conflict. Equally, all the others must be coached in their role as defenders of the proposals so that the 'devil's advocate' is not subjected to personal attacks.

Since most companies have a hierarchical structure, many junior employees feel uncomfortable arguing with or 'attacking' a more senior person and may well not be able to make a full contribution. This gives rise to the need for a chairman.

The Chairman

Each team meeting needs to have a clearly selected and appointed chairman. Initially this is likely to be you until you are comfortable with everyone's ability to play their roles satisfactorily. Once this point is reached, you should rotate the chairmanship amongst all the team members.

The role of the chairman is to ensure that everyone in the team is able to make their contribution and actually does so. It is also to ensure that everyone remains focused on the objectives of that meeting.

SELECTING THE CQI TEAM – TEAM BUILDING

It is unlikely that anyone would expect to pick a new team and then see them perform at a high level immediately – the CQI team is no exception. They will need a detailed briefing on the objectives of the programme, the background to continuous quality improvement, and the work done in the research phase. They will also need briefing on the proposed methodology that the team will use to carry out the strategic and tactical planning work.

Initially, there is likely to be a healthy degree of scepticism about the whole process and you will have to sell CQI to the team – this is a big challenge and your first in terms of convincing others that what is being undertaken is for the benefit of the company and everyone working there. Now is *not* the time to bring in the senior management – it is your team and you have got to be convincing.

In the words of the managing director, this is how one company, which I'll call 'Acme Wonder Widgets', tackled this first team meeting:

'George had been the driving force behind the whole idea – he's the human resources director – and was godfathering it with the board. He had selected Caroline, one of his first line managers, as the project leader.

'The first CQI team meeting involved about nine people but I was not invited. When I asked about this, Caroline explained that she wanted to handle the first meeting without the pressure of my presence. This seemed reasonable. George was going to be there so I had no worries. Normally, any business or project meeting takes place in our meeting rooms at head office but Caroline, in a break with tradition, requested permission to take the entire team to a hotel for a two-day meeting over a weekend. This was a radical departure but, since George supported the move, I agreed.'

Peter, a shop floor union official and one of the team, takes up the story: 'I think most of us were reluctant to give up our weekend but Caroline was persuasive and the hotel, in North Wales, was really nice. We got there on the Friday evening and right from arrival Caroline took control and welcomed us and really made us feel part of a team.

'That evening after dinner we held our first meeting at which Carline outlined what we were going to do. There was a lot of resistance and, I have to say, I was leading the charge. But she didn't back off nor did she lose her temper and get flustered – which was what we expected given she's only a kid.' (Caroline was 28 years old and had worked for the company for three years; Peter was 54 and had been on the shop floor since leaving school at 16.)

'That first meeting was really tough on her. We were all giving her a hard time but she kept going and she never asked Mr Green (George) to help her out. It was impressive and won her a lot of respect.

'Eventually we all started to listen to what she was saying – anyway the long and the short of it is, we none of us got to bed before three in the morning.'

Caroline picks up the story: 'That first meeting was just about the worst experience of my life. But I'd selected the team and I was determined that they were going to come with me on the CQI trail – I could have wept when Peter and the others started in on me but eventully we had an understanding. But the thing that told me I had won was the next morning – I was down early and was standing on the terrace wondering how the day was going to work out when Peter joined me and put forward some proposals for the first working session. I quickly rethought my plans and incorporated his suggestions.'

George Green, the programme's godfather: 'After the rough ride on the Friday night I was pleased to see Caroline and Peter together and, although they can still crash into each other over policy on occasions, they are the ones who drive the whole thing. The rest of the weekend went very well indeed – it was a great idea to hold the meeting away from the office as it stopped an awful lot of posturing and allowed the team building activity to work. At the end of the weekend we had a CQI Team and a CQI Team Leader in the person of Caroline.'

This is not an unusual approach – a number of people have reported to me that building the team before launching the programme has been critical to the success of the project. Getting away from the work environment puts all the participants on neutral ground and allows the team leader to control the process more. In some companies off-site meetings may be common occurrences, but for many team members – especially those selected from among the workforce – the idea of a conference at the company's expense at a reasonable hotel is a powerful team building tool.

The primary objective of this first meeting is to make the team members feel part of a special project group – something that is highly motivational and motivation is going to be needed if the programme is to be a success. I suggest that you focus the first meeting on team building: the introduction of the CQI ideas and plans can then be carried out in a much more receptive climate. This is not to say that you should spend all the time on team building activity – you must sell them the CQI programme, but this is easier once the team spirit is in place.

You will notice I say 'sell them the CQI programme' – it is vital that all the team buy in to the programme and become enthusiastic promoters of it. It is no use telling them about CQI: they have to become committed.

The CQI Master Plan – First Steps

As soon as you have your team in place you will need to build on their commitment by getting them involved in the preparation of the strategic master plan. And one of the best places to start is by establishing the feasibility of the goals. It may seem strange to suggest that the goals you have decided on may not be feasible, but all too often you will hear complaints that the goals are unrealistic, that they are unachievable. And, I have to say, all too often the people saying these things are right. Goal setting is a skill and, as with all skills, has to be learned – you cannot just go out and set goals without some practice.

At this stage in the CQI planning process goal setting can only be done against benchmarks and against progress – ultimate and specific goals are likely to be unrealistic unless based on something concrete. Benchmarks are available from a number of sources – previous experience, experience of others, best practice, and calculation, to name but four.

In Chapter Three we discussed the objectives for the CQI programme and we selected a range of management behaviours (Goals, Feedback and Recognition, Training, Service Quality, etc.), internal relationships, and key financial ratios. In Chapter Four we looked at how we could measure them to determine where we are now. But we need to go further – we need to decide what our position will be after a period of time. 'But,' I can hear you say, 'CQI is about continuous improvement and therefore there can be no end point.' Quite right – but we can set incremental goals based on where we are now.

Let's look at some examples of incremental goals;

- The dimension score for service quality might show a 48% positive, 22% neutral and 30% negative score. A one-year goal may be to improve the score by 5 percentage points – that is, to achieve 53% positive and 25% negative as a minimun position. In this example, it is important to note that a more positive score requires an upward move in the positive score AND a downward move in the negative score.
- The Net Revenue per FTE (i.e., the net income to the company divided by the number of full time employees) might show £30,000/FTE. Your one-year goal may be to achieve 5% increase (i.e., to a minimum of £31,500/FTE).
- You may have a 'best practice' benchmark to put in place a structured process for annual appraisals. A one-year goal could be to have the process developed and tested and all the training needs identified.

Do you have to have the results of the surveys to carry out a feasibility check on the goals? Not really. Let me explain: in the first example above the incremental objective or short-term goal was to make the scores 5 percentage points more positive, not to achieve a fixed percentage of 53%; in the second example it was a 5% increase, not a fixed value of £31,500/FTE. Now, the question is: is 5% improvement a feasible, realistic, achievable goal for a soft benchmark dimension and is a 5% increase in Net Revenue per FTE a feasible goal inside one year?

To a certain extent, only you can answer that question for your own company, but a 5–10% improvement in any score in a year should be an entirely achievable result and I like to see incremental improvements of this size being undertaken. On this basis, a score should improve by 30–60% over a period of five years. As an action, therefore, you should establish what percentage improvement you wish to set as a goal for each of the dimensions and key financial ratios – irrespective of what the actual scores for these dimensions and ratios might be.

In the case of 'best practice' standards, once you have selected which ones to apply to your company you will also know whether or not you have already achieved them. (Your survey of the best practice standards is a single question on each standard: 'Have we achieved this standard – Yes or No?') It is reasonably straightforward to establish what can be achieved towards the standard in a year.

Actually, it is not unrealistic to set goals that require all best practice standards to be achieved within two years – indeed, the only reason for taking that long is that developing the processes for the standards takes time and then you have to train everyone and make sure the standards are being applied.

One company decided that they would benchmark their business against best practice standards and ignore internal benchmarking. They listed and described all the standards in a manual which they issued to all the staff with instructions to apply them properly. The management were very proud of their achievement and described their best practice standards in glowing terms. It will come as no surprise to you to learn that a simple check on the standards showed that 58% of the processes and procedures in the manual were not applied at all (because no training was given and no follow-up organised by management) and the rest were applied in a haphazard manner. Imposing procedures and then not ensuring they are carried out is not the way to achieve CQI.

Your team will, of course, be very involved in this process. Working in small groups, they will be deciding on and checking the short-term goals and will also be involved in a more detailed analysis of the company in terms of the best practice standards and the key financial ratios.

You, on the other hand, should be developing the principal milestones for the whole CQI programme.

THE CQI MASTER PLAN – MILESTONES AND DEVELOPMENT CURVE

The programme milestones are those things that have to be achieved along the way to full implementation of CQI. They should be established so that you can measure how far along the route you have travelled. My starting point for this is to break down the programme into phases and then establish milestones for each phase.

Phase I Commitment and Development

1.1 Senior management buy-in and personal commitment to the programme

1.2 CQI programme leader identified to be responsible for the implementation of the programme

1.3 Resources formally dedicated to the CQI programme

1.4 CQI team identified and recruited

1.5 Benchmarking indicators established

1.6 Benchmarking research (surveys, etc.) carried out

1.7 Establishment of feasible goals for incremental benchmark improvements

1.8 Development of CQI Master Plan

Phase II Involvement And Assimilation

2.1 Formal CQI awareness and accountability is pushed down through all management levels

2.2 All managers have CQI goals and objectives

2.3 Senior management are personally involved in highly visible and meaningful CQI activities

2.4 CQI communications strategy and tactics fully implemented

2.5 Action plans for all CQI dimensions are established and actively implemented

2.6 Quality circles and quality teams are established in all departments

2.7 Comprehensive quality goals are established for each work team and a monitoring process established with monthly reporting

2.8 CQI team have visited and trained, where necessary, all departments within the company

Phase III Maturity

3.1 Re-benchmarking surveys have taken place after 18 months covering all dimensions

3.2 New CQI incremental goals have been established

3.3 The results of the re-benchmarking have been communicated to all members of management and the workforce by senior management in group meetings

3.4 Action plans have been reviewed, modified and implemented

3.5 Comprehensive quality goals in each team are reviewed, modified and re-established

3.6 All managers now have CQI goals as part of their personal objectives

3.7 CQI 'Quality Awards' are established and the criteria made known to everyone – first awards made

3.8 Best practice standards have been fully implemented and reviewed for completeness and application

Phase IV Maintenance

4.1 All milestones in Phases I – III have been fully met

4.2 Re-benchmarking surveys are run every 18 months, results disseminated, and the new CQI incremental goals established

4.3 New action plans developed and implemented

4.4 CQI is now a way of life – new CQI tools are continually developed and used, the programme is self-perpetuating, the CQI team has been reduced to an oversight group, formal CQI reviews take place and new budgets are allocated as necessary

The milestones listed above are the essential ones and you should supplement them with others that reflect your own particular programme and its objectives. In general terms, the milestones should be grouped under four headings – Statesmanship, Efficient Tools, Quality Professionalism, and Standards and Measures. The links between these four groups of milestones and the phases are shown below.

Let me make a small digression. The concept of Statesmanship in the CQI programme is very important. All of us, and especially those in senior positions, need to demonstrate statesmanship: we need to be the ambassadors of the programme and of the whole concept. This means we have to manage in a CQI way, run our business lives on CQI principles, and demonstrate our commitment to CQI at all times. Statesmanship must be encouraged and the best way to do this is to encourage all senior managers to play an active and regular part in the programme, get them to talk about CQI at every meeting, get them to ask for all business reports to show how CQI has been implemented and how CQI principles are involved. But remember, it's not just the seniors who have to show statesmanship – everyone must do so and awards for statesmanship should be a frequent and regular part of the recognition process.

Once you have the milestones, you need an effective monitoring process to check on progress and to act as a motivational tool for the business especially during the early parts of the programme. A large international bank developed the idea of a 'learning curve' – a graphic representation of their progress. I have taken this idea one step further and developed the CQI Development Curve – to recognise that we are not just learning things but are developing a whole new way of approaching the business and its survival.

Measuring performance on the Development Curve requires you to weigh the value of the Phase I and II milestones so that when all have been achieved you can mark the curve at 80%. The remaining milestones then represent the remaining 20%.

As far as timing goes, it should be possible to reach the end of Phase II after 18 months and the end of Phase III after a total of 3 years. Thereafter, the programme enters the maintenance phase which has an infinite length. This time scale is realistic in my experience – some of my clients have taken longer but most have achieved significant improvement within this time frame. One client was so successful that the business, which covered 13 countries and involved around 150 people, was well into the maturity phase within 18 months. The original prediction of three years was reduced to 27 months and I found my involvement was no longer required.

Once you have developed the milestones and your CQI

	Commitment & Development	Involvement & Assimilation	Maturity	Maintenance
Statesmanship	– Business specific – Dedication of resources	⟵ TOTAL CORPORATE COMMITMENT EXEMPLARY LEADERSHIP ⟶		
Efficient tools	⟵ BASELINE DATA SYSTEMS – Develop Plans	MIS SYSTEMS MARKETING	EQUIPMENT QUALITY SYSTEMS – Refine & automate	⟶ – Update to meet challenges
Quality Professionalism	– Establish clear roles for employees – Develop training programme	– Managers accountable for goals – All managers trained	– Staff accountable for goals – All staff trained	– Process updated and reviewed – Training programme renewed and updated
Standards & Measures	– Determine initial benchmarks – Set initial standards	– Refine all standards and measures	– Coordination of all internal & external benchmarks	– Periodic reviews of standards. – Updated measures for the business

The CQI Milestones Matrix. The matrix shows the links between the milestone groups and the phases in the programme.

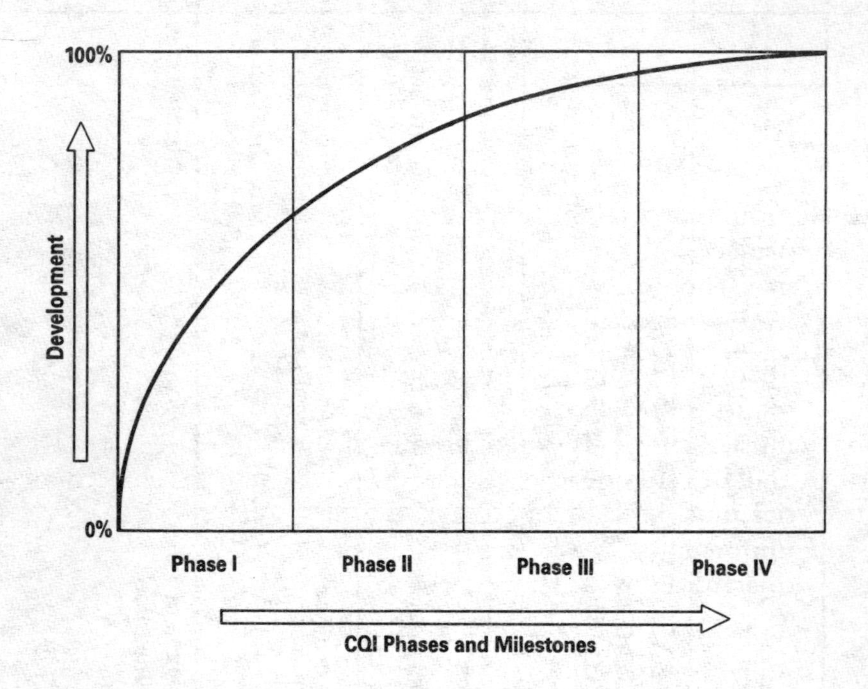

The CQI Development Curve. Rapid progress towards full implementation of a CQI programme can be achieved in the first phase and by the end of Phase II around 80% of the implementation has been reached. The end of Phase III represents 95% on the Development Curve.

planning team are beginning to pull together the Master Plan, you should focus on one vital strategy without which the whole process will come to nothing. I am, of course, talking about the communications strategy which is itself the first step towards obtaining the final go-ahead and commitment from senior management and the buy-in of the workforce.

SUMMARY POINTS

- The preparation of the CQI Master Plan is a team effort.
- Although you will remain the overall leader of the CQI project, since you own the objective and the task, you may choose to appoint other members of the team to lead certain activities.
- Team leadership is a function – a job that has to be done – and there are three critical areas on which a leader must focus: Achieving the TASK. Building the TEAM, and Developing the INDIVIDUAL. These three activities must be kept in balance, but it is vital that a leader understands that it is his or her primary responsibility to achieve the task.
- As leader you need to pick a team that has the right skills, competencies, and attitudes to achieve the task – including willingness to think the unthinkable, challenge the status quo, question everything and refuse to accept answers just because they are based on received wisdom.
- The strategic planning team should have around seven members selected from across the whole workforce. It should include one senior manager and one middle manager; the rest should reflect the mix of the workforce.
- To get the best from the planning process the team needs certain defined roles to be played: Godfather – to protect the programme and to represent it to the senior management, Rapporteur – to keep the records of the project and to present the results to management and workforce, Devil's Advocate – to challenge every proposal to test its validity, and Chairman – to ensure that everyone has made a contribution to the process.

- Before starting work on the planning, the CQI team needs to be made into a team – in this way they will become committed to the process and will champion it all the time.
- CQI has to be *sold* to the team and not just *told* to them if you are to get the best from them.
- The first step for the team is to check the feasibility of the goals that have been set and to establish realistic, achievable, and measurable goals for the incremental improvements that are the basis of CQI.
- As CQI leader, you must develop the Milestones for the programme so that everyone can see how they are developing and how the process is going. Measure performance in terms of the Milestones and display this by using the CQI Development Curve.

6

Communication and Commitment

Communication – image – presentation – medium • budgeting • middle managers – the key to success • selling CQI to the staff

COMMUNICATION

THE COMMUNICATIONS STRATEGY is, perhaps, the single most important element of the CQI programme – get this right and you should have a successful programme, but get it wrong and there is little chance of the programme delivering the benefits of which it is capable. The communications strategy is the process through which you sell the benefits of the programme to everyone in the company. If they buy the benefits, then they'll buy the programme. And if they buy the programme, then they will be committed to its success.

Marketing people will tell you that a successful communications strategy is based on a strong brand image – an image that everyone immediately associates with the subject. They will also tell you that presentation is everything. To these two key points I add content, which is critical, and the medium to be used.

The Image

I once made the mistake of assuming that the programme image was not important – not a mistake I will make again. I firmly believe that you need to brand the programme clearly so that everyone instantly recognises that what they are seeing is important and is about CQI.

Let me illustrate this point.

Acme Wonder Widgets, which we met earlier, was set in its ways – solidly based around old-fashioned values and with a very traditional approach to most things – but had recognised the need for 'quality improvement'. Their original approach was heavy and ponderous and, although perhaps acceptable to the board members, it was not likely to be well received by the rest of the workforce. Caroline, the CQI project leader, decided to work with her marketing colleagues to establish a fresher, more up-to-date approach; one that would attract attention but still convey a quality image.

In the process of developing the branding Caroline and the marketing team decided on a consistent typeface, font size and layout for all front covers as well as specifying paper quality and colour. You may well say that this was going over the top, and that is your decision, but these things play an important part in the way the document is accepted. I don't for one minute suggest that glossy brochures and colour printing are required and whether to use coloured paper is a choice that depends on corporate culture as much as anything else – however, you should take as much care over the CQI communications materials as the marketing director takes over product brochures.

The consistent application of the brand image to all things connected to the programme soon generates acceptance – it also demonstrates commitment by the company to the process of change. The brand image has to be appropriate and easily recognisable. And I advise against the use of humour in the branding since this can confuse and irritate the reader, and trivialise the content.

The Quality Process
as applied to
Acme Wonder Widgets

The full strategic plan for
turning the business into
a quality company and to
promote profitability

CQI
Profitability
through Quality

The Way Forward

A Quality Initiative by
Acme Wonder Widgets

The front covers to the CQI Strategic Plan for Acme Wonder Widgets. The original version (L) shows a very traditional approach and the revised version (R) shows branding – the 'logo' was used on all CQI publications and the 'footer' appeared on all front covers.

Presentation and Content

Presentation and Content are inextricably interlinked and which comes first is driven by how, where, and to whom the message is to be delivered.

A client once told me that when presenting to the senior group executive I would have five minutes of his attention and if I had not made an impact in that time he would cease to listen. I decided, therefore, on three overhead slides with four bullet points on each which I would support with a short commentary. However, I had other slides available and a big thick report for background reference.

The client had called me in to examine the possibility of

implementing a CQI programme and the presentation subject was the state of the business and what should be done. The slides are shown on page 99.

The group executive did not particularly like the message I was delivering, but I did have his attention for the next 90 minutes. Being concerned about the 'political' impact of my message would not, at this point, have served any purpose – my job was to tell the management what it needed to hear and not what it may have wanted to hear. It was time to show leadership in the project and I would have been letting the CQI team down if I had not shown commitment and confidence.

I am a great believer in the use of overhead slides with bullet points and graphics on them and I seldom make business proposals without them. However, your company may be one in which greater use is made of flip-charts (which can always be prepared in advance) or it may have a culture in which visual aids are hardly ever used. Even if this is the case, I still recommend the use of overhead slides: the introduction of a new way of making presentations generally helps the overall success of the communications strategy. Furthermore, overhead slides are vital if presenting to groups.

The general rule is **keep it simple**. What you put on the screen is an *aide-mémoire* to those who are listening to your presentation and should make key points. You should then support each point with a well-thought-out commentary. As a guide, work on the principle that you use one slide every two to three minutes unless the slide is just a summary. Make no more than four or five points per slide and use lots of graphs and diagrams on the principle that 'a picture is worth a thousand words'. Finally, and perhaps controversially, I suggest you limit your presentations to just twenty minutes for most audiences and make them even shorter the more senior the audience. Senior managers are very busy people and frequently have a great many things that have to be dealt with in a meeting – your presentation on CQI is just one of them. However, if you

Where are we now?

* Low overall staff satisfaction
  ➢48% positive 34% negative
* High staff turnover
  ➢23% leave per annum and 38% want to
* Declining market share
  ➢down from 14% to 11% in 24 months
* Declining net revenue
  ➢down 10 % over same period last year

Options

* Do nothing and continue the decline
  ➢projected survival time 3 years
* Cut costs and reposition the business
  ➢projected survival time 5 years
* Undertake CQI – with total commitment
  ➢projected survival time forever
* Pray
  ➢projected survival time unknown

CQI – a strategy for survival

* Improve staff satisfaction
  ➢and improve staff retention
* Improve management behaviours
  ➢to improve productivity & performance
* Improve internal quality
  ➢and improve customer satisfaction
* Improved staff and customer satisfaction plus improved productivity & performance will result in improved net revenue

Three slides used to make an impact. The strategy was to identify key issues of relevance to the audience (in this case, the group executive) and then show how they could be addressed.

need twenty minutes then take it, otherwise you will not be doing your job.

If a report is the best way of presenting your information then think it through carefully and make sure you have it structured logically – there are a number of books that can help you on this subject (for example: *Writing that Works* by Roman & Raphaelson, *Getting Through* by Howard, and pages 66–77 of my book *Managing for Performance*). My own preference is to supply an 'executive summary' at the beginning of the report (to tell readers what you are going to tell them) supported by the body of the report (tell them) and ending with an overview summary of what has been said (tell them what you told them). I generally include all graphics and data in the body of the report rather than use appendices or annexes. Whatever approach you use, try to be consistent throughout the communications process as this makes comprehension easier.

Finally, let me stress one thing about content and presentation – it must be **relevant** to the audience: relevant to their work, to the company, and to the objectives of the programme. Something that is deemed irrelevant will very quickly find its way into the waste-bin.

The Medium

There is an ideal medium for every message and you should not be limited in your use. The most successful communications strategies I have experienced within CQI programmes have made full use of

- overhead slide presentations
- reports
- newsletters
- staff newspapers
- videos

- discussion groups such as quality circles
- meetings (especially involving senior managers talking to the workforce)
- questionnaires and other feedback processes.

The distribution of each item to the intended audience includes the internal mail system, fax, e-mail, hand delivery, staff meetings – whatever method will ensure that the message reaches all the intended recipients.

But what should go into the CQI communications and information strategy?

The purpose of the communications strategy is to obtain commitment to the programme and maintain that commitment. But it is a bit of a chicken-and-egg situation as to whether the communications strategy comes before the commitment or once it is in place. A friend who does most of my communications work suggests that obtaining the commitment should be the first objective and the rest of the strategy can be built from there.

Her normal starting point is to circulate a short newsletter that briefly explains the idea behind the CQI programme. She then holds a series of meetings to discuss real quality issues so that people gain an understanding of the benefits *to them* of a quality programme. At no point does she go into any great detail as to why a quality programme is good for the company, she just focuses on how it benefits the immediate audience.

Each meeting is followed up by a report in the quarterly CQI newsletter and she always makes certain that people are mentioned by name – everyone likes to see their name in print. Where the budget permits, photographs are included.

After the initial round of meetings, the strategy switches towards the use of newsletters and staff newspapers – again with people mentioned by name and lots of photographs. Managers are also provided with 'management briefings' covering activities in their areas. At a more senior management level, the strategy focuses on keeping people informed through 'quality updates' and presentations at management team meetings.

The purpose of the communications strategy is to build and maintain commitment to the CQI programme and the most important audience, and the one that needs to be addressed next, is the senior management of the company. Many senior management teams recognise the need for quality improvements which they see as a competitive advantage and a tool for achieving greater profitability – for this reason you will probably have found a good deal of lip-service being paid to CQI. This attitude will probably have gone as far as to allow you to establish the CQI team and undertake the initial benchmarking activity – the cost of doing these is relatively low in budgetary terms – but this must not be taken as giving full commitment to the long-term process of CQI.

By this stage you will have completed a Strategic Master Plan, see page 103, and you will know how you wish to proceed based on the plan and the communications strategy you have developed. You also have all the elements upon which to base the budget calculations that will have to be done before you can go to your senior management and ask for their commitment and the necessary funds.

BUDGETING THE CQI PROGRAMME

Your budget calculations have to be based on two distinct elements:

- What will the programme cost in terms of cash and other resources (human resources, overheads, support, etc.)?
- What financial benefits will the programme bring?

These two elements will allow you to calculate the return

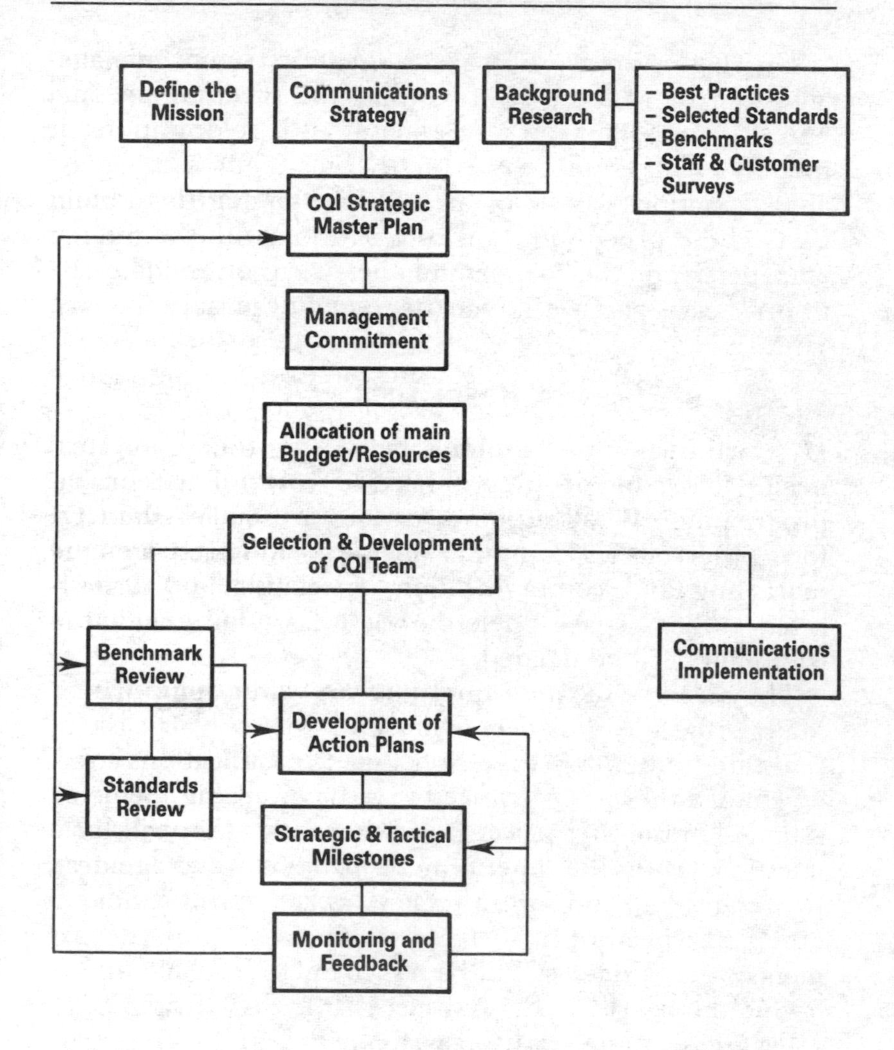

The summary flow chart of the CQI Strategic Master Plan developed by Acme Wonder Widgets

on the investment in a way that today's management understand – the effect on the bottom line. At this stage you cannot afford to rely on the obvious but intangible benefits of improved customer relations, customer retention, and greater staff and customer satisfaction: you will have to show the financial benefits as well.

The first step is to make a qualified and educated 'guesstimate' of the cost of running the programme. And here I genuinely mean an estimate – it is seldom possible to put an absolute figure on the cost, but some good approximations can be obtained by looking at the various parts of the process and, in co-operation with the finance managers and other department heads, you should be able to price each part with a fair degree of accuracy.

Resources

My starting point has always been with the resources I need. How many people will need to work full time on the programme? What support services are required in the form of secretaries, computer operators and programmers, marketing and communications specialists? What dedicated office space, telephone lines, fax machines, computers will be needed?

The answers to these questions are often illuminating. Many project leaders overestimate in this area to be on the safe side, and when they find that the budget is revised downward they are convinced that they were right to overestimate in the first place. But if they had approached the questions with the intention of producing an accurate figure and had supported their estimate with the views of others (the head of finance and the heads of other departments), the budget would be less likely to be challenged.

One approach used very successfully by Chris Steel, a colleague of mine, was to carry out the 'global' resources review and budget and then to show how and where this could be funded. The advantage of this, he found, was that while department heads resist a new budget being allocated through fear that their own would be cut, many of them were prepared to fund resources through existing budgets spread across the whole company.

In one such situation: 'We calculated that the resources required for the programme amounted to £100,000 per annum and there was no way the Board would have approved this over the objections of the department heads. I approached each of the three major departments – financial control, marketing, and human resources – and suggested that the costs could be amortised by "borrowing" people and resources from them on an as-needed basis. Financial control were quite happy with this since it meant a reduced budget allocation and they could claim that, since we were going to be using some of their people for the MIS, they were making a contribution. Marketing were happy because I asked to be allowed to operate from their office – which they saw as a logical extension of their activity – and it would cut down on the substantial commitment they would otherwise have had to make in terms of people and time. And human resources were happy since they had a couple of under-used secretaries and by borrowing them I did not need increased FTE. Finally, when everyone realised that no additional people were required and that we would be using existing facilities and resources on an *ad hoc* basis, the board signed off on the budget of £45,000 (my costs) and felt that they had saved over £55,000.'

Of course, you will have to make your own budget calculations but if you follow Chris Steel's approach you may well find that the correct use of resources (an objective of CQI!) means that the additional budget requirements will be low. Very few of my clients have more than two people dedicated to CQI and most have only one – everyone else is seconded on an *ad hoc* basis to everyone's satisfaction.

Direct Costs

The direct costs of the programme are those funds needed for CQI activities – off-site meetings, travel, the communications strategy, the benchmarking surveys and the outside consultancy services you will need in connection with them. These can generally be estimated quite accurately and need to form a committed budget. Exactly what you decide to include will be decided by the way you want to run the programme, but CQI is about quality improvements and not about having a good time at the company's expense. It must also deliver tangible financial benefits if it is to find acceptance amongst management.

My approach has always been to be very honest about what I would like to do but always to split the direct costs into two parts – essential and desirable. In any fight over the budget the parts that get cut are always in the desirable but non-essential costs.

Investment Return

With the resources and direct costs of the programme properly calculated you now know how much the CQI programme will cost. The next step is to establish some approximations as to the financial benefits of the programme – since all managements have to know the approximate return on their investment, the CQI programme should be positioned in their minds as an investment in the future of the company.

Where do you start? My choice is to go for the easy one first and look at the key financial ratios we are using as benchmarks. In Chapter Five I suggested that key financial ratio benchmarks should be set to improve 5% over the ratio for the previous year – this is straightforward compound growth and it is easy enough to calculate the financial return, especially if you are using Net Revenue per FTE as a key ratio (of course, in doing your projection you have to assume that the FTE will remain constant for the period).

Other key financial ratios will also provide a guide to the investment return, but you will also have to look at some of the intangible benefits that come from improvements in the soft benchmarks. For example, the savings in training and recruitment if staff turnover was reduced by 5%; or the reduction in acquisition expense if customer satisfaction was improved and repeat business increased by 10%. And what about the increased revenue that could be generated if productivity and performance of each and every individual in the company were to go up by 10% as a direct result of managing better and establishing performance enhancing goals.

In each case you must assume that no re-engineering takes place and that the number of full time employees (FTE) remains the same – in this way your projections are based on similarities of starting and finishing positions.

This is a fascinating exercise and the resulting investment return could be enormous – Acme Wonder Widgets, when they did this calculation, came up with a figure of in excess of £1,500,000 in the first year against a programme cost of £125,000 per annum. On that basis the investment return in the first year would pay for the programme for 10 years. Did they achieve it? Yes, and more. Their net investment return was around £1,000,000 in the first year rising to £2,125,000 in year three.

Results such as these are not uncommon – a banking client attributed a 50% increase in net revenue from customers over one year to the 'Performance Management' activities they undertook to improve their soft benchmark scores and you will hear similar stories from any company that has seriously addressed CQI issues. Whatever the actual financial outcome of the programme, the investment return over a three-year period is likely to be in the region of 10 times the programme costs.

While improved financial ratios and large investment returns are not *in themselves* the key reason for carrying out CQI, they are powerful incentives for any Board or management team and a very useful 'sales' tool for you in obtaining management commitment to a CQI programme and to achieve the resources and budget needed. The cost-benefit analysis is compelling.

Credits and Debits

It is wise to present the budget data in as clear a manner as possible and that means you should be prepared to discuss the debit side of the equation as well as the credit side. On the credit side there are the increased productivity and performance that you identified and 'valued', the increased revenue and profits which the key financial ratios will identify, and the potential cost savings.

Cost savings are an emotive as well as financial issue – over and over again you will see in the newspaper and on the television how companies are pursuing that alchemist's dream – normally through cutting jobs among the workforce rather than among the management.

Management repeats the mantra of the 1980s over and over again: 'we must cut costs if we are to survive'. This is a dangerous myth, an illusion which managers the world over want to believe as they struggle for ways of securing the future of their company and their own jobs. All too few of them understand that cutting costs is only possible where cost controls have failed and managers (themselves) have failed to utilise their resources correctly. It is a purely *defensive* strategy. If those same managers were doing their jobs properly, then the resources in the company (budgetary, human, etc.) would all be utilised to maximum advantage and there would be no excess costs to cut unless they also wanted to cut revenue. Over the years I have met all too many managers who can tell me where the next £1,000,000 of cost savings are going to come from but not where the next £1,000,000 of revenue growth will be found.

If we recognise that we have failed in our responsibility to utilise our resources correctly, our first action must be to reverse the situation – use our resources to generate more revenue. **Cutting costs is an admission of failure as a manager and you must discourage your colleagues from thinking this way**.

All this is not to say that cost savings are not possible or desirable. If we can radically improve the way a process is carried out (through re-engineering) and can produce 'orders of magnitude' improvements in productivity as a result, then we will generate cost savings *on that process*. But those cost savings only become real if we can, *at the same time*, either dramatically increase the productivity of the individuals involved or reassign them to another (new?!) revenue generating activity.

On the debit side of the equation you should look at the costs involved as a result of the disruption that change can produce, the costs associated with the introduction of new processes, and the potential for short-term declines in productivity as the workforce is retrained and the new processes learned. To quantify the cost of disruption and the introduction of new processes is very difficult – not least because we have yet to identify and determine the new processes – but you should, at least, make your management aware that such downside potential is likely but short-term and that each action should be viewed as a separate investment issue.

Time

Just as the CQI programme cannot be done on a shoestring, it does not have to cost a lot. Similarly, it cannot be done by just one person, it has to be a team effort. And, most importantly, it is not a quick-fix solution to a problem, it is a long-term solution to the need for a survival strategy.

Earlier we talked about the programme taking around three years to reach maturity in terms of the programme milestones – although, of course, the CQI cycle must run for ever. Some senior managers, unfortunately, have a short-term perspective and if the programme is not to be completed in 'the current budget cycle' (12 months, 2 years, etc.) then they are unlikely to support it. Unfortunately, to quote a Chinese proverb, 'something built in a day is unlikely to last a thousand years' and CQI cannot function effectively if it is to be treated as 'flavour-of-the-month' and forgotten next year. CQI is not the latest short-term panacea.

You must, therefore, make it very clear that CQI is a long-term project that will require funding over a long period of time although the greater part of the cost is in the first three years. You should also stress that most of the investment return is likely over a three-to-five-year period. This is illustrated in the graph below, which represents data for Acme Wonder Widgets.

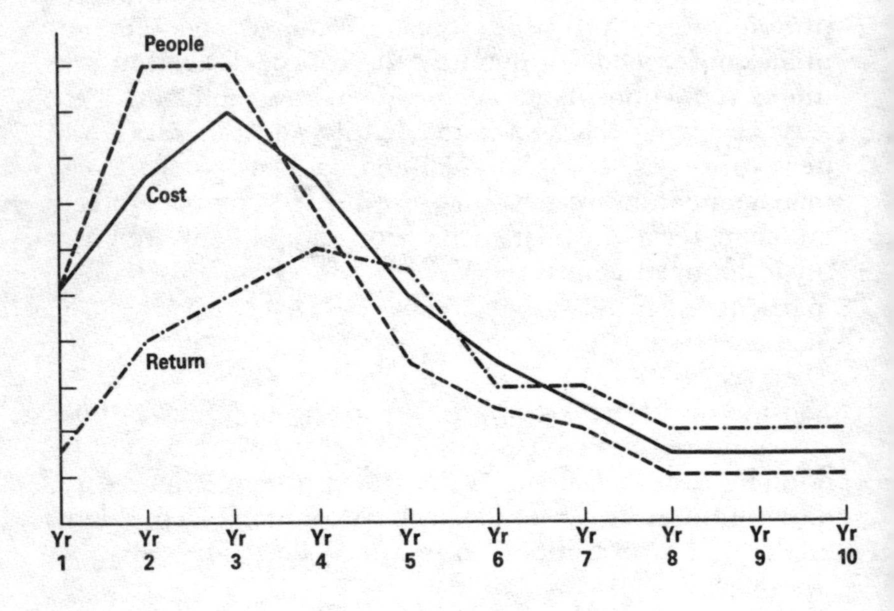

The distribution over time of the costs, the investment return, and the need for people for the CQI project.

Armed with all your budget information, the communications programme, the CQI plan and a good slide show, tackling the problem of obtaining board or senior management commitment is unlikely to be a problem and with this secure you can move on to selling the programme to the rest of the workforce.

MIDDLE MANAGERS – THE KEY TO SUCCESS

Over and over again I have found that getting the middle management on the side of the CQI programme is a vital step. These are the people who really control and manage the workforce and without their full support nothing gets done, not a wheel turns, no work is turned out. But get them behind a project and anything is possible.

Why are they so vital?

Because of the way that most companies are structured, managers decide who is to do what, when, and how. Managers are, therefore, in a position to dictate the activity levels of a business and how their people will react to or participate in a programme. You need the middle managers on your side if the CQI programme is to succeed, and yet they see themselves as the people with most to lose as a result of change. Forget this fact and you are almost certainly doomed to fail.

Some years ago I was on assignment with an international brokerage. My job was to set up a European office and I had been working on this for about 18 months and was basically fulfilling the role of branch manager until a permanent person could be found. Things were going along quite well when suddenly head office started requesting more information from the individual staff members without going through the managers. After a while the branch managers (myself included) became rather suspicious; we started talking to each other about what was going on and it soon transpired that head office had decided to restructure the company and to set in place some worthwhile quality changes – but they had failed to communicate this plan to anyone and seemed to want to go over the heads of the branch managers.

At one of our regular branch manager meetings we raised the subject with the head of the brokerage and were more or less told that the process was a 'done deal' and that we should make sure our people worked within the new structure. I don't think there was a single happy manager in the room – we all had a great deal of autonomy (after all, we worked in different parts of the globe, never saw anyone from head office and seldom met the boss), and we each ran our office in line with local laws and practices. Now we were being told that everything should change. At no time did the senior management attempt to get the branch managers on their side and the resulting breakdown in communications was almost tangible. Within nine months of that meeting all the branch managers had left the company and a once thriving brokerage of more than 100 people had shrunk to a core group of 20 people all reporting to head office.

It need not have been like that. If the senior managers had obtained the backing of the branch managers before launching the programme, the message would have been enthusiastically sold to the teams and the business would have become a major force in the marketplace – instead, it withered away and the partnership broke up, all due to senior management having alienated the middle management level. It was almost as though they were scared of the impact of the plan. Because they had shrunk from the possibility of conflict they failed to communicate and destroyed their business in the process.

My preferred approach is to work closely with the middle managers and involve them very early on. They are the ones who really understand the fundamental issues of quality and often know how and where to focus to solve the problems. Their knowledge is invaluable and the more they are involved the greater their commitment to finding solutions – this leads to a growing commitment to the programme itself. Once you involve this group of managers in the programme you will find that they make staff and time available to assist you. The fact that they may have a hidden agenda for doing this (perhaps they think that commitment and support for the programme will reflect well on them) is of little importance at this stage:

whatever their reasons for assisting, their support is valuable.

To get them 'on side' requires the selling of the benefits of the programme as well as handling their very real – if imagined – concerns that they will lose influence, power, budget, FTE or whatever. Such concerns cannot be dismissed as irrelevant and you would be wise to explore them in a very open manner.

The approach of Caroline at Acme Wonder Widgets is a useful example of what to do. Caroline held group meetings – a total of 15 of them – with all the middle managers. She was accompanied by George Green, the programme's godfather, and the purpose was to explain the benefits to the managers of participating in the programme and to identify their fears and suspicions. Typically, managers worried about their jobs being at risk, other concerns focused on declining power, reduction in budgets, and the possibility of radical changes in the way their departments would work.

In each case Caroline and George were very open. They acknowledged that all these things were possible, while stressing that *the objective was not to reduce staff* but *to make better use of the resources available*. Over and over again, they emphasised that reassignment was always a possibility but that the vast majority of managers would stay in their current roles, often with enhanced responsibilities and powers. At no time did they promise, or even imply, that any manager would be unaffected by the programme – in fact, they promised quite the opposite: that every manager would be affected by the move to CQI and that *CQI was a survival strategy* that could ensure that everyone had a job in the future. At the end of the meetings the managers felt that the programme leaders had displayed openness and honesty; although they were still concerned, they believed they could trust Caroline and George and that the CQI programme was not as threatening as they first thought.

Selling CQI to middle managers is not easy, but they are the single most influential group of opinion formers in the company and time invested in gaining their commitment is time well spent. I cannot stress enough that they are the key to the successful implementation of CQI.

SELLING CQI TO THE STAFF

Using the top-down cascade approach, the next step is to sell the programme to the staff. This tends to be less of a problem because most staff see very quickly that any programme designed to improve the way they carry out their work can only be a benefit to them. They are also the ones who have first-hand experience and knowledge of the quality problems and they are the ones who really know how to fix them.

You should arrange that each team leader or section manager holds a team meeting to explain the purpose of the CQI programme and how it will involve and affect the team in terms of their commitment to the programme and its likely outcomes. You, of course, should also attend these meetings to answer the technical questions and outline the company policy, but the main presentation should be by the team's own manager.

To aid the communication process you should provide the manager with a slide presentation and hand-outs to ensure the right message gets across. In addition you should have talked the manager through the presentation and identified where you will need to assist. Finally, you should have worked with the manager to establish what the next steps to be taken are as far as he or she and the team are concerned.

The whole process needs a top-down **and** bottom-up approach that draws in everyone at all levels.

Again, time spent selling the programme is time well spent, but be careful not to raise expectations that cannot be fulfilled as that is a quick way to lose credibility and the commitment of your people.

Once the programme has been sold to everyone in the

company and their commitment obtained, you can then move on to the detailed development of the action plans.

SUMMARY POINTS

- Communication is the most important strategy within the CQI programme and the process by which the benefits of the programme are sold to everyone in the company. It is the process through which commitment to the programme is obtained.
- A strong 'brand image' is necessary within the communications strategy and care should be taken to ensure that a clean, appropriate image is chosen so that anything 'branded' is associated with CQI.
- Presentation and Content are interlinked. The presentation should always be consistent with the brand image of CQI but is governed by the content.
- Select the presentation to suit the content **and** the audience. Do not be afraid of breaking with the company presentation style if another style would get the message across better.
- The use of overhead slides (sometimes called viewfoils) is the simplest and most effective way of presenting information to groups. A well-thought-out set of slides will give both presenter and audience a feeling of comfort and security even if the message is difficult.
- Select the medium for each message carefully. Use all media appropriately and include overhead slides, reports, newsletters, staff newspapers, videos, discussion groups, meetings and questionnaires.
- Commitment to CQI has to be supported by dedicated resources – people, money and equipment.

- The CQI budget should be presented in terms of all the costs involved and all the financial benefits that could and should accrue.
- Always budget carefully and show all the costs involved. Show where internal costs can be obtained – not all costs have to be provided by additional cash budgets, some can be met by 'borrowing' resources from other departments.
- Remember to include the direct costs of the programme – such as off-site meetings, travel, the communications strategy, and outside consultancy costs associated with the benchmarking surveys.
- Identify the investment returns – if these are not obvious then work on the principle of 5% improvement per annum in the financial indicators, and the likely cost savings that will result from improved soft benchmark dimensions such as decreased staff turnover and increased productivity.
- Be clear that there are credits and debits associated with the programme – identify the likely benefits of the programme but ensure that the debit side is equally understood; disruption, short-turn drops in productivity while re-training and re-engineering takes place, and the like.
- Ensure that everyone understands that CQI takes time to produce benefits and that costs will peak some time before the investment returns – often by two years or more – but that the investment returns are likely to be orders of magnitude greater than the costs.
- Make sure senior management understands that the CQI programme will have to be funded, albeit on a decreasing scale, over a very long period of time.
- Once Board level and senior management commitment has been obtained it is vital to sell the programme to the

middle managers. Middle managers must be involved and committed as they can ensure the success of the programme or, by lack of commitment, its failure.

- Cascade the programme – get the middle managers to sell the programme to the staff and support them while they do it.
- Get the 'workers' involved at all levels.

7

Action Planning

Principal areas of focus • Human resources • training • management

UNTIL NOW we have been focusing on getting the CQI programme established. We have looked at the background to such programmes, how we establish where we are now, and how to obtain the commitment necessary for us to proceed. Once that commitment is given by senior management, it is down to you to make things happen.

To do this you will need your team since the amount of work involved will be beyond one person's capacity if it is to be delivered within an acceptable time. The best way of handling the workload is to assemble a working group to prepare the necessary action plans for each of the principal areas. These groups will then act as the main implementation mechanism. The composition of each working group will be dictated by the area in question – in the human resource area you need people from HR, in the marketing area you need people from Marketing – but in each case one member of the group should be from your main CQI Team. This person will act as the conduit between the working group and the members of the CQI team and help to ensure that their plan fits with the others and no conflicts arise.

With the working groups in position they start by asking four interdependent questions:

- What are we doing now?
- Is it necessary?
- How can we do things better?
- How can we do things differently?

The answers to these questions will provide you with the action plans that you need to drive the programme forward. The CQI action plans are the very visible aspect of the programme and the one that actually provides the mechanism to achieve the changes that are being looked for.

What is so important about these questions? Well, they tackle mind-set – the way we think about situations. We all go through life with minds that are trained by our experiences and we apply this knowledge when we analyse a problem and try to produce a solution – this is our mind-set and it can lead us to all sorts of false conclusions depending on the situation and the relevance of our experience in dealing with it. But it goes further – mind-set also results from the conditioning we have received and the culture in which we operate. Expressions such as 'we don't do things that way around here' or 'in this company we do things differently' or 'as an Englishman (Frenchman, German, Italian, Greek, Chinese, etc.) I would approach that in a certain way' are frequently heard whenever people meet to resolve issues.

Mind-set can severely limit our approach to issues, it can stop us analysing the situation correctly and it can stop us finding the most appropriate solution. If Albert Einstein had allowed himself to be limited by the mind-set of the scientific establishment of the day he would never have concluded that $E=mc^2$ and powered flight would never have been achieved if conventional wisdom (mind-set) had guided the Wright brothers.

To break free of the existing mind-set it is necessary to think differently, to eliminate preconceptions as to what is 'correct' or 'possible', and refuse to accept limits. It means asking not just 'can we do this process better?' but 'do we need to do this process at all and if so, why?' and 'if we need to do this process, can we do it differently?'

Doing things *better* and doing things *differently* are frequently not the same thing at all. 'Better' presupposes that we improve the current process; 'differently' suggests we change the process completely but achieve the same outcome – albeit in a vastly improved way. And that's the key – if we need to carry out a process then doing it differently must produce *orders of magnitude improvement* in terms of productivity (i.e., 200% faster, 75% less expensively, with 50% fewer people, etc.) and if that is not the case then doing it differently is a waste of time and effort.

In one re-engineering process, the man in charge claimed proudly that as a result of his actions he had saved the company well over two man-years – this sounds very impressive until we realise that two man-years represented a saving of 2% in human resource terms. Why did they bother? If he had saved 20%, 40% or 60% it would have been worth doing.

Many people find 'thinking differently' very difficult because there are few national or corporate cultures that encourage their people to challenge the status quo. Such people are frequently difficult to manage and disruptive to the smooth running of the organisation. Furthermore, there is a lot at stake – people's power, prestige, status, job and income are all on the line and challenges to these are not likely to be welcomed. However, thinking differently must be done.

So how do you go about it?

The starting point is to go back to fundamentals. Ask yourself 'why are we in business?' and 'what do we need to do to meet that objective?' Andrew Carnegie, a man who understood the fundamental purpose of business, defined the purpose of business as making money: 'our objective is to make money by making steel'. Unfortunately, many people forget or never understand the basic purpose of business. Some think that it is to give the customer what he wants, to make things, or even to provide the workforce with employment. All these are outcomes but they are not the reason for the business, which is to make money.

Some years ago I sat through a slick presentation by the chairman of a large public company at its AGM. He insisted that the objective of his company was to provide customers, in a service-orientated way, with the latest, technologically sophisticated, superbly engineered, competitively priced, widget. The presentation was wonderful and, for a while, we all believed him. He was impressive, but he came to a juddering halt when one of the shareholders stood up and said, 'Mr Chairman, we admire your sales message but are you going to make money for your shareholders?' There was a stunned silence then we all applauded, which quite upset the chairman.

If you keep in mind that the objective of your company is to make money you can then move on to how to do it. From this you can determine what needs to be done in a variety of areas to ensure that the ultimate objective is achieved.

Principal Areas of Focus

The areas on which you need to focus can be divided into three groups – the 'soft' or behavioural, the marketing, and the operational:

- Human Resourses
- Training
- Management

- Internal Communications
- Marketing
- Sales

- Support Administration
- Business Processes
- Production and Operations

I will look at the three 'soft' or behavioural areas now and the rest in the next chapter.

HUMAN RESOURCES

During the earlier research stage described in Chapter Four you will have identified a number of key benchmarking standards that you wish to achieve and you will have measured the company against these criteria – have you or have you not achieved them already? Now is the time to start the process of achieving those that have not yet been reached.

The key standards to be reached by any quality company should include a comprehensive human resources strategy covering, at the very least:

- full descriptions for all job functions
- career paths for all functions
- training commitments for all functions
- comprehensive objective (goal) setting processes
- performance review processes
- performance appraisal processes

The most important objective in the HR field should be to have a properly planned HR strategy. This is obvious but all too often overlooked. Many companies adopt the approach that the only HR strategy needed is one that ensures that people are recruited when needed. This is very short-sighted given the ever increasing value of our human resources – people are not some tool to be picked up and abandoned at will, they are an educated, skilled, and highly mobile resource that walks out of our business each night and, we hope, returns the next day.

Job descriptions

The first requirement for your HR strategy is to have a proper understanding of what jobs are necessary for the company to reach its objectives. I am not talking about the jobs currently being done but about those **jobs that must be done** if the company is to survive. And by jobs, I mean the job function and not the positions that have to be filled nor the number of people required. For example, in a retail outlet such as a shop the job functions are likely to include:

- stock selection
- stock ordering
- stock reception
- stock storage
- stock pricing
- shelf filling
- serving the customer
- receiving payment and giving change
- record keeping
- payment of suppliers
- maintenance of premises (cleaning etc.)

Have I missed anything? Possibly, and I am sure that you may be able to identify other functions. One you are likely to put your finger on is management, but here I would disagree with you – management is not a job function that is necessary in a retail outlet such as a shop. If you have selected the right people to carry out all the job functions and trained them properly, given them the responsibility and the authority to carry it out then there should be no need for a management function.

Now, you may strongly disagree with what I have just said but remember: think differently and ask yourself, 'What added value to the shop does the manager bring – what increased revenue is generated?' Then ask: 'What would happen if there was no manager?' Would the shop cease to function, would it fail in its objective which is to deliver the goods to the customers in a way they want, at a price they can afford and at which the shop makes adequate profit?

I raised this question in a management seminar and was told that the manager was vital – he or she had to make decisions about pricing, about staffing levels, about stock, about opening hours, and a myriad other things. I agree all these things have to be done but they are all sub-sets of the job functions listed on page 123. The delegates argued amongst themselves for some time and eventually, very reluctantly, agreed that the function of management was not necessary.

I am frequently surprised at the variety and range of jobs that can be excluded under the criteria of 'necessary for the survival of the company' or 'necessary for the company to reach its objectives'. For example: many levels of management and especially senior management are unnecessary – although *parts of their current function* are necessary. I am not saying that the management function is not a *desirable* one to have within the company – leaders are required, some decisions cannot be made at lower levels, and someone has to take ultimate responsibility for the success or failure of the business to reach its objectives – but it is not a *necessary* one in terms of survival.

Once you have identified all the job functions in your company, the next step is to write a detailed description of the function in terms of what has to be achieved and the minimum performance standards (MPS) required. For example: shelf filling may have, as part of its description, the following statement: shelves must be checked at the beginning of business and immediately after lunch and any stock item that is below the pre-set minimum levels should be replenished to the pre-set maximum level within one hour. Such a description must be detailed for each and every part of the job function.

The objective of this job function description is to establish exactly what is required to fulfil the role, but it does not state that a *person* has to carry out the function. Deciding whether the job should be done by a human or a machine is something that is affected by a number of other factors which are addressed, along with deciding on the number of people required, when looking at business processes. At this stage we are only interested in the job function itself.

Having completed the job function descriptions for all the necessary jobs, you should then turn to all the ancillary job functions that are thought to be desirable – managers, secretaries, receptionists, clerks, drivers, team leaders, and so on. Each of these job functions also needs a full description along with minimum performance standards.

Finally, working from the description of each job function you will need to develop a skills, competencies and attitudes profile for the job. This profile defines the skills needed for the job to be done to the minimum performance standards and the range of additional competencies you want – for example: team working skills, task focus, self-confidence. It will also describe the sort of attitudes desired – loyalty, customer focus, and so on.

Clearly, identifying, describing, and profiling all the job functions in a medium to large company is a massive undertaking and is probably well beyond the capacity of one person to achieve within a short space of time. So set milestones in place within the action plan so that the work gets done in clear stages within certain time frames. If it is going to take a year to complete this work, then milestones that state that all necessary job functions have to be identified within four months and fully described within seven months will ensure you can monitor the performance of the work group against its action plan.

Career paths

With the job descriptions complete, it is possible to develop career paths for each function. This is simple where functions are clearly delineated with a hierarchical structure – junior clerk, senior clerk, supervisor, section leader, etc. – but, in many of today's companies such structures have been eroded or abandoned and a flatter, less clearly hierarchical structure is in place. Despite that, or even because of it, you must establish career paths for your people since everyone needs an idea of where they are going. Without this sense of progress, money in increasing amounts becomes the only way that they can measure their career development.

In some companies, career paths are not an option. I worked closely with one company in which people were recruited for specific jobs and the chance of movement within the firm was remote. In another, a huge computer company, people were even recruited on fixed-term contracts at the end of which they had to compete for a new contract. This short-term approach had certain advantages in that it meant that staff could be reduced almost at will and the overall long-term employment costs kept to a minimum. The disadvantages, on the other hand, were obvious: staff with no long-term expectations were seldom committed to the work and rarely produced above 60% of their capacity. Their quality of work was low and they spent a great deal of the last six to nine months of their contract looking for the next job – which they frequently found outside the company. Staff turnover was an extraordinary 58% amongst all levels up to department head and the training bills were enormous.

While it was not appropriate to return to the paternalistic jobs-for-life approach, the company did develop a more structured approach: short-term appointments were abolished and replaced with 10-year contracts for the generalists and project-based contracts for the programmers. The financial effect was felt almost immediately: staff turnover dropped, productivity increased dramatically, goals were introduced, business revenue increased and the overall HR costs were reduced by 40% within two years – yet they still employed the same number of people.

Career paths and structured career development may not be possible or even desirable in your company, but it is important to provide clear guidance as to the options available to change jobs within the company and to be transferred to other departments. Those options must be real and must be accompanied by the necessary conditions to ensure that people transferring to a new job are actually suited to that job and have the right skills and competencies.

In a recent survey for one of our major clients we found that dissatisfaction with the career structure was a major contributory factor to an overall high level of staff turnover.

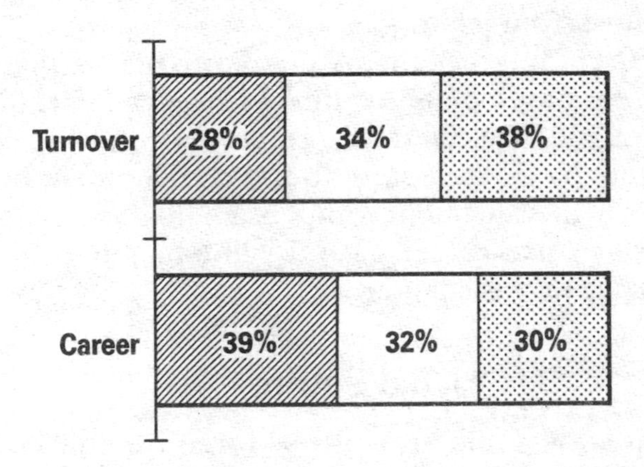

Turnover (above) shows that 28% of the respondents responded positively to the statement 'If I could, I would like to leave the company within the next 12 months'. Career (below) shows that 30% of respondents in the same survey indicated they were dissatisfied with the opportunities for career development.

Training

Going hand-in-hand with career development is the subject of structured training programmes that are directly related to the job functions. Unfortunately, whenever there is a call for cost savings, the first budget item that gets pruned is training. This is, very simply, false economy. Unless the company has an expert staff selection and recruitment process permanently in position or is prepared to pay outside headhunters or executive search agencies to recruit even their lower level staff, then it is almost impossible to recruit and retain people who have the skills required. Almost always a new recruit will require a good deal of training and, as each new level of challenge is reached, new training requirements come into play. It is vital, therefore, to ensure that your company has identified the skills and competencies required in each job function and that suitable training programmes have been established to support them. These training programmes then become 'training commitments' or prerequisites for people as they move from job to job.

One outstanding benefit of this approach is that even where career development is limited by structure, the fact that training is available and required is often seen as a commitment by the company to the development of their employees and can often compensate for restrictions in other areas such as career path. I will look more closely at training later in this chapter.

Goal setting

Given that most members of a workforce are unlikely to be delivering the best possible performance of which they are capable, it is essential to establish a comprehensive goal setting process – a process that encourages, even demands, continuous improvement in performance. Here we are talking about performance-enhancing goals, not the minimum performance standards that are part of the job description. For some years now companies have paid lip-service to MBOs (Management By Objectives) and have confused these with goals. It is important that you ensure that a clear differentiation is made between the two – MBOs are part of the job description, goals are about enhancing performance.

Many managers believe they know how to set goals – one manager claimed that 'all you have to do is decide what you want the person to do, increase it by 25%, and tell them that is their new goal'. Unsurprisingly, none of his team achieved their goals – he had forgotten (or never knew) the basic rules of goal setting: they must be 'owned' by the individual, they must be SMART – specific, measurable, accepted, realistic, timely – and they must take into account current conditions, performance and ability. This manager needed a crash course in goal setting and I am sure you will come across similar managers among your colleagues.

So what's to be done? Your HR action plan will need to include specific training for managers on how to set goals and how to monitor them. And **every** manager should undertake the goal setting module so that a consistent approach is used. At the very least this module should teach the concept of SMART goals designed to achieve measurable improvements in performance on a continuous basis. There is little point in establishing a system in which goals are accepted by the workforce, but they only achieve the same as before. You need to get managers and workers to think in terms of quarterly (or, better still, monthly) goals that require a 5–10% improvement each quarter. One thing is certain, annual goals are virtually valueless in terms of motivation and continuous improvement because people just cannot keep focused on a target that is so far away.

In *Managing for Performance*, I outline the whole process of goals and goal setting as it is a key part of good people management.

Goals require monitoring and people require feedback

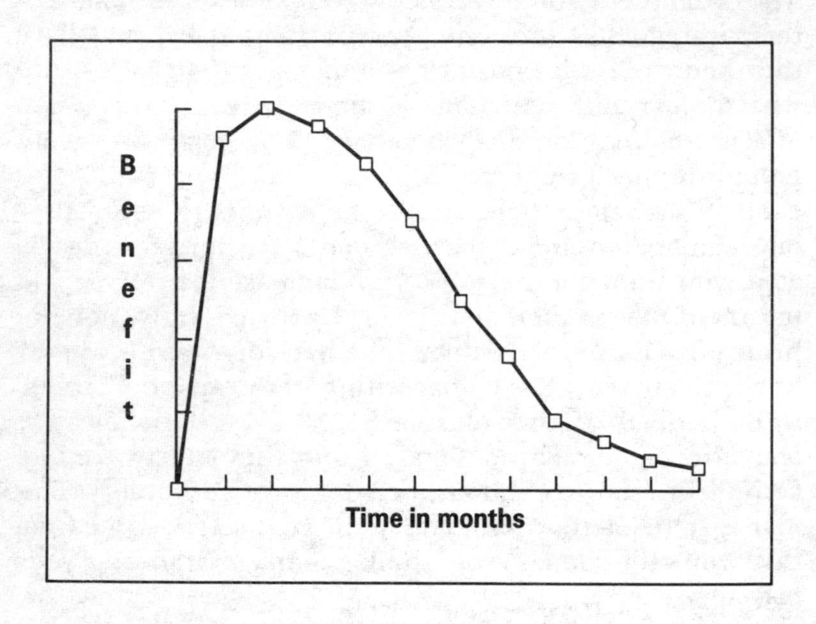

The declining benefit of goals over time

on their performance against their goals and these are the next key issues which you have to address.

Monitoring

Telling people how the business is doing or how the department is doing is fine if the only goals that exist are for the business or department, but it is practically valueless if the goals are, as they should be, individual or team-based. People need to know explicitly how they are doing against their own goals. Monitoring is only useful if it is continuously available to the person whose performance is being monitored. The airline pilot's main goal is to keep the aircraft a certain distance above the ground and to fly a certain course. To check how he is doing he has an altimeter to tell him how high he is above the ground and navigation equipment to tell him where he is and in what direction he is flying. It would be of little help to the pilot if the only people who could see the altimeter and navigation equipment were the passengers – and rather unnerving for the passengers if they knew the pilot did not know how high he was or in what direction he was flying.

Your action plan should seek to establish performance monitoring processes for each group or individual that has goals. These processes should be as simple as possible – one big failing of modern MIS systems, the computer generated information and data reports, is that there is too much information on them. The performance monitoring process must be clear, simple, and directly related to activity and appropriate for the person or group involved. It should not be confused with the information reports that teams and individuals use to control and monitor the process or business.

It is probable that you will meet resistance to the introduction of simple monitoring processes since many managers, in particular, seem to believe the more information they supply the better. A banking client in the UK supplied a weekly report for each and every account executive that covered around 20 to 30 lines of information ranging from new accounts opened through the average balance per existing account to the revenue per relationship – each report ran to about twenty or thirty pages, depending on the number of accounts handled. The managers believed that this report was suitable for monitoring the performance of the account executives in terms of their goals and that all the information was necessary. When we checked with the individual account executives we found that they were using just three lines that directly related to their goals and only referred to the other data at month end to ensure that the business process and their portfolio were meeting expectations. After a considerable battle the managers finally agreed to provide the three or four lines used on a weekly basis and to produce a full report once a month. The benefit of the monitoring process was maintained, but the saving in time and paper was enormous.

My advice is to establish a simple paper-based monitoring process directly related to the goal cycle – if the goal is to achieve a target by the end of the quarter, then the monitoring process should provide weekly or monthly information covering only what has to be achieved that quarter. For example: if a widget maker needs to produce 10,000 widgets in the quarter the monitoring process should show the target (10,000) and how many have been produced since the beginning of the quarter. Don't fall into the trap of expecting the individual to remember a starting point other than zero – each quarter's goal should be complete in itself.

Finally, do not forget the importance of the motivational effect of goals. Goals are personal objectives aimed at improving performance and the monitoring process must reflect this. Monitoring of performance should be a public rather than purely private process – we all like to see how we are performing in comparison to our colleagues and the best way to show this is to display performance publicly but in a way that will enhance, rather than damage, motivation. Some individuals and teams will be less experienced than others or are working with different constraints and will, therefore, have different goals from their colleagues. To avoid demotivation setting in when they realise that their goal may be much bigger or smaller than someone else's, you should establish the public monitoring in terms of **percentage** of goals achieved. After all, everyone in the group knows that for the group to reach its goal then everyone in the group must achieve their goal **no matter what size that goal is**. Therefore, seeing how well people are doing as a percentage of their goal is motivational to themselves and to others – it also helps build team spirit.

I like to use bar charts for most monitoring processes: the one below shows the performance of a sales team.

The use of this type of public monitoring has been found to be the key to both motivation and the continuous improvement of the individual's performance.

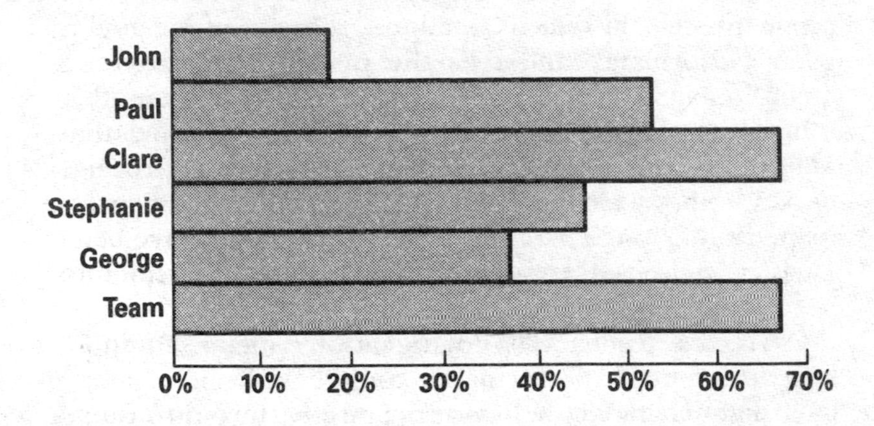

A performance monitoring chart showing the performance of a sales team against their quarterly goals. The performance is recorded as a percentage of their goal.

Feedback

Of course, supplying an accurate monitoring process is the first step in providing feedback as to how people are doing. But feedback requires more, and you will need to decide the best feedback process to use to ensure that continuous improvement is achieved. You will then need to build this into the action plan so that it is implemented consistently and correctly.

Feedback consists of a number of parts:

- performance feedback from monitoring
- positive feedback
- corrective feedback
- performance reviews (not to be confused with performance appraisals).

The feedback provided by performance monitoring is generally passive and carried out through the process outlined above. Positive feedback, on the other hand, is an active process in which someone recognises a piece of good performance and tells the person concerned in a quick, public and appropriate manner. It can range from a hand on the shoulder and a 'well done' to a properly applied process whereby the person providing the feedback specifies exactly what was done well and how it has helped towards the overall objective. This sort of feedback must be delivered as soon as a praiseworthy action has been identified.

Corrective feedback is **not** negative criticism. It shows how an acceptable performance could have been enhanced by a slight change in behaviour. The structure is to recognise the performance as good, identify what could have been better, supply a little bit of coaching, and convey 'keep up the good work'. If the behavioural change needs to be more significant, then coaching has to be used rather than straight feedback.

Carrying out positive and corrective feedback is a skill that most managers will be unfamiliar with – most have been exposed to the attitude that 'we only talk to our people about their performance when it is not acceptable'. This prevalent but destructive approach has no place in a CQI company and you will have to include training and coaching for managers, at all levels, so that the required skills can be acquired and used daily. You will also have to establish a 'quality control' process to ensure that the skills are being applied in a consistent and appropriate manner.

Performance review

The final step in providing feedback is the performance review. Let me be quite clear: performance reviews are not the same thing as performance appraisals, which I will deal with next. Performance reviews are a tool for motivating your people and should take place whenever a team member has produced a performance **below** or **well above expectations** and they are **always** non-disciplinary.

The purpose of a performance review is to determine the cause and reasoning behind someone's actions or performance *before* we use further skills to coach, train, or manage the situation. The use of reviews when performance is below expectations is obvious, but they should also be used when performance exceeds expectations by a significant factor – for example, if someone's performance results in exceeding their goals by 20% or more – as this may mean that the goal setting process has not functioned as it should.

The performance review is generally an informal process involving a relaxed interview, in private, between a manager and a staff member. The interview provides a forum for:

- reviewing progress
- giving more detailed feedback
- identifying strengths and weaknesses
- resolving problems
- establishing roles and responsibilities
- defining standards to be achieved
- setting priorities and goals
- identifying training and development needs.

As with other forms of feedback, conducting performance reviews is a skill that most managers will have to learn and specific training programmes will have to be established as part of the overall action plan.

Performance appraisal

The subject of performance appraisals is one in which much controversy and misunderstanding occurs. Until now we have been looking at good management *behaviours* whereas performance appraisals fall into the category of good management *processes* and really form part of the human resource management structure.

Performance appraisals, frequently referred to as 'annual appraisals', fulfil a very specific purpose. They are to:

- review, objectively, the performance of the individual against the job function responsibilities and minimum performance standards
- determine whether the performance is sufficiently above the minimum performance standards to attract a salary increase, bonus, or promotion
- establish the development goals for the coming year
- establish the career moves the individual would like.

Performance or annual appraisals are vital activities that allow a company to check that it does have the right people doing the right jobs and to confirm the match between the individual's actions, skills and competencies and the requirements of the job. The formal appraisal report also monitors that the development of the people is going according to plan and helps the HR department plan career moves and job allocations.

Unfortunately, many companies confuse the function of the performance review with that of the annual appraisal: consequently neither is done well. This can be overcome by establishing the performance review process through training and monitoring and by adopting a slightly more mechanistic approach to appraisal: after all, it is possible to determine whether the person has or has not fulfilled the responsibilities and has or has not met the minimum performance standards. Two other approaches I have also found extremely beneficial are to have the individual concerned complete a self-appraisal which is then reviewed by HR along with the manager's appraisal, which should also be signed and agreed by the individual, and to have each manager appraised by the people who report to him or her. The first ensures that a clearer picture of the individual in relation to the job is obtained and the second ensures that managers are delivering quality management, which is as important as fulfilling their responsibilities as far as more senior management is concerned.

Establishing a proper appraisal system is a strategically important move in the CQI process, but it is one that requires strict application and this has to be monitored to ensure that all parts of the process are carried out within a very short period of time (say 1–2 months) at the end of the year. My own preference is to ask individuals to carry out their self-appraisal and the appraisal of their manager in the last month of the cycle and to have all the management appraisals completed by the end of the first month of the new cycle so that reassignments and training programmes can be implemented before the end of the first quarter.

Training

The training action plan is a direct derivative of the Human Resources action plan and it is, therefore, necessary to have completed that before trying to develop a plan for training. Once again we are looking at two key issues: best practice and strategic impact.

It would be best practice if your company had a comprehensive training programme attached to each job function. Such training would fall into three main categories:

- knowledge
- skills
- coaching.

Although we should not forget the whole subject of personal development, it is a secondary training issue that can generally be addressed through the provision of self-directed programmes covering time management, negotiating, self-awareness and other such subjects. These self-directed learning programmes should be made available to all staff through audiotapes, video productions, printed materials, computer-based programmes and through access to open seminars run by outside training companies. If your company is working in foreign markets you might also like to provide self-directed language courses. If it is at all possible your training department should build up a comprehensive library of such material with strict rules as to how long a programme can be borrowed by an individual. They should also keep good records of which staff have undertaken which self-directed programmes as this will provide a guide to which of your people are interested in their own development and prepared to make some effort of their own – in itself a guide to ambition and attitude towards career development.

Training is an *investment* in the future of the business, **not** an *overhead*.

Knowledge training

When you develop the job function descriptions you will also decide the knowledge necessary to fulfil that function. Here we are talking about specialised knowledge, not the general educational background which you will also have defined but which should 'come with the applicant' when they apply for the job or are appointed. It would not be commercially wise to try to educate people except in the specialised knowledge that the job demands. For example: bank tellers need to be numerate and literate and should not be taken on if they are not – but they also need to understand banking rules, how to accept deposits and make outgoing payments, which forms to fill in for which transaction and a thousand and one things which they can only learn from their employer.

Unfortunately, teaching people what they need to know to do the job is something many companies are very bad at – they seem to believe that they can get by with their staff learning 'on the job'. This is very short-sighted: an untrained staff member needs continuous supervision, is unlikely to be working effectively (perhaps as low as 20% of minimum performance standards), will make mistakes which cost money to rectify and is, in general, an ineffective and expensive overhead until such time as they have the right knowledge. Imagine what the cost would be, in financial and human terms, if airlines trained their pilots like that – it hardly bears thinking about! If new employees receive comprehensive knowledge training before taking up their role they are likely to be working at 60–80% of minimum performance standards straightaway, are unlikely to make costly mistakes and need much less supervision – thus they are a cost-effective addition to the team almost immediately.

This imbalance between training off the job and learning on the job comes about because many companies see training as an overhead, a cost. But training is an investment, since it is designed to provide us with an effective resource that will help in the overall generation of revenue and profits. Your training action plan should focus on ensuring that the right knowledge for each job is conveyed through specialised knowledge training courses and is provided either before the person takes up the job function or, at the very latest, within 2 or 3 months of starting. One of the findings that crops up repeatedly in staff surveys is that the right training is not provided in a timely or appropriate manner: this is both a 'best practice' and a strategic issue.

Each job function should have a structured training programme attached to it and only when all the relevant knowledge training has been successfully undertaken can people be said to be fully trained in that function and ready to move up to the next level within that function. A well-structured human resource strategy will ensure that promotion or grade changes will be as firmly linked to completion of the requisite training courses as it is to meeting minimum performance standards and exceeding performance-enhancing goals.

Generally speaking, knowledge training is best run in-house by qualified trainers pulled from 'the line'. They should be coached in their role as trainers and used when necessary before rejoining the business. This means that the only overhead is really that of developing the training course itself – something that should be done by an experienced trainer in conjunction with the line management in that particular area. In this way the cost of the training function or department is kept to a minimum. If your company is too small to have a specialised training department, then you will need to bring in an outside consultancy or training provider to help you develop the courses and to run them for you.

But, you may say, what knowledge training is necessary for routine 'labouring' jobs such as secretaries, cleaners, machinists, fork-lift-truck operators and so on? Think about it for a moment. All these functions require a knowledge of their machines, how the company operates, how their job fits in to the overall business activity, what their reporting line is, where and how to get supplies, and a number of other pieces of information if they are to be effective. Their training may only take a day or perhaps two or three days, and there may only be one training course for them in that function, but you should still ensure that the right knowledge training is given.

Skills training

The skills training programme is also an important strategic tool and, as with the knowledge programme, is a vital investment in the future of the company. Here you would need to review the competencies and skills needed for each job function and then develop the necessary training courses. In many cases you will be able to handle skills training within the business but in the case of the 'soft' or behavioural skills, such as sales, advanced negotiating, public speaking and presentation, telephone techniques and so on, you may well decide it is easier and more cost-effective to buy in the training from one of the specialist providers. If you do take this route you need to ensure that the courses are customised to suit your business environment and your industry.

In many ways skills training, especially for management, is the one area of investment that is likely to produce the greatest return. Behaviour modifications in the way we manage and the way we interact with others (especially our customers) have a dramatic effect on the effectiveness of the business as a whole. One of my banking clients is a classic example of this: they used my training programme, *The Performance Management Workshop*, to train all their branch managers in performance management techniques and skills and then ensured the new approach was rigorously applied. The result, as far as they were concerned, was a $10 million or 40% increase in revenue and vastly improved staff satisfaction and effectiveness.

In the more practical work areas such as production, operations, administration and even marketing it is very likely that you can make the necessary skills a prerequisite for appointment to the job, leaving little in the way of practical skills training to be done. However, refresher courses are always useful and necessary to ensure that everyone's skills are right up to date and that nobody has picked up bad habits. You will probably find that much of the refresher training can be done by coaching and thus reduce the cost to just the initial investment in establishing the programmes.

Coaching

Coaching is the process of imparting knowledge and skills to others during day-to-day work. In this way it is very different from knowledge and skills training which has to be done off the job and, in many cases, away from the work environment in facilities designated for training purposes. Coaching is on the job training and requires particular skills.

Recent research amongst leading companies shows that all of them place coaching at the top of the management activities necessary for achieving success and maintaining a leading advantage. Sports stars and teams all achieve their success under the guidance of coaches who are responsible for improving their performance and maintaining their fitness, technique, and winning ability. Yet few companies make coaching a core management activity and many of them have no coaching programme of any description. Common responses include: 'of course we coach our people if they ask for it' and 'if someone is not delivering then the managers have a word with them.' Worse, the people who say these things believe that what they are doing is delivering coaching!

Coaching is not something that is done on a one-off basis – it is a **continuous programme of activity** designed to ensure that our people develop and enhance their skills and use them effectively to the benefit of the company. Coaching is also every manager's prime responsibility in people management terms. It helps motivate, it enhances performance and effectiveness, it improves revenue generation and reduces costs, and it costs nothing. Unfortunately, the vast majority of managers are very bad at coaching and few see it as an important part of their job: this has to be changed.

Your coaching action plan must, therefore, first focus on training your managers in coaching skills and in how to implement a coaching programme. While this is being done you can develop the actual programme and communicate it to the workforce.

Coaching and coaching skills can be taught and once all managers have been adequately trained further coaching courses are unlikely to be needed on an on-going basis. Moreover, coaching is a generic activity – the skills used are the same whatever the actual activity involved – so I suggest you make use of one of the outside training providers. This is certainly the most cost-effective method of addressing this issue. What is rather harder, however, is ensuring that the coaching is really being done. It is no use just stating that it has to be done, you will have to establish some minimum performance standards (MPS) within the manager's job function description and then check on the performance by way of feedback from that manager's team.

The coaching programme should include a minimum of one day's coaching with each individual every quarter – one day a month would be far better, but that might be impracticable. Everybody in the team should be involved and not just those whose performance has fallen below expectations – it is better to reinforce good technique and behaviour than try to correct deterioration. It is certainly more cost-effective.

Once everyone is aware that coaching is a serious and vital investment in the future of the company it will probably become a routine activity, but it must be checked on at regular intervals – coaching is an intrinsic part of continuous quality improvement and that requires it to be a continuous management activity.

At the more senior levels of management it is often difficult to convince people of the continuing need for everyone (from the Chairman downwards) to coach their immediate staff. This is where you will need to work on the senior management team and use the CQI godfather to make sure the activity is in place – perhaps by stressing that if they are seen to be coaching then those below them will realise that it is part of their job and the process will be implemented via a cascade. Coaching at this level is less likely to be so structured but it still involves a more senior manager spending a day with the people who report directly to him or her, and observing how and what they do and advising on how they could improve their performance.

Senior managers coaching those below them is also a useful communications technique and can be motivational for all levels of staff. They see that senior managers are interested in what happens throughout the organisation, despite the common complaint that 'the bosses' do not know what happens below them, do not know who performs well and who does not, and never listen to the 'other ranks'.

One critical area of coaching is the communication necessary before the initial implementation. If your company has no formal coaching programme the introduction of one will be viewed with suspicion – why is everyone suddenly so interested in it? Are they looking for people to get rid of?

From my own experience of introducing coaching programmes, there are two stages to the communication process: (i) a company-wide briefing on what the coaching programme is about, how it works, and what its purpose is, and (ii) a direct briefing to the managers on how to introduce the programme to their staff and how to make it work.

The first step is best achieved through the CQI Newsletter. Take plenty of space and explain the programme in detail, starting with the benefits of coaching – improved personal performance, enhanced skills, motivation, better communications, greater likelihood of achieving objectives, and so on. Then describe the coaching process itself as discussed above. Finally explain that coaching is to be an official policy, all managers and supervisors will have coaching goals relating to their teams, and everyone will be asked to report on the coaching they receive. This last point helps to bolt the policy down and will help ensure that it really happens.

The second stage is best accomplished during a one- or two-day coaching workshop during which all managers and supervisors are instructed in the techniques and skills of coaching and how to set up a coaching programme. They will also have plenty of opportunity to practise the skills off-the-job before returning to put them into operation.

Management

Your management action plan must focus on the implementation of best practices and that means

(i) managing people,

(ii) making correct use of resources,

(iii) making profits, and

(iv) ensuring the long-term survival of the company.

Unfortunately, in many companies the management focus is on processes, not people. They misuse resources by not keeping costs under control. They try to save money by getting rid of people rather than using them better to make more profit. They fail to achieve their targets in terms of revenues and costs. They think short-term – frequently only one year ahead.

Changing bad management practices is not easy and you will need all the support that you can get from senior management and from the board of directors if you are to achieve your objective – which should be to make all managers directly responsible and accountable for the performance of their team and with sanctions in place and used. This last point is the key – for too long companies have paid lip-service to managerial goals but have seldom, if ever, applied sanctions for failure. Over and over again you will hear stories of managers who have failed to reach their individual and/or team targets, yet still pick up bonuses based on overall company performance. Excuses are always found. If a team has failed to deliver against target the person to blame is the person responsible for the goal – the team leader or manager – and that person can only be excused if the failure is due to something totally beyond their control or something that cannot be foreseen and planned for. Failure as a result of the incorrect use of resources, not leading the team properly, or not planning carefully enough cannot be excused and the manager or team leader should be penalised – perhaps even fired.

Why should you take such a strong line? The answer lies in the performance of management in a wide variety of companies over the last ten years. Under pressure from institutional shareholders, who require regular dividends from, and growth in, their investments, management has focused more and more on short-term performance and has frequently used the share price as the principal way of monitoring their performance. This kind of management generally looks at the current year's performance and the plans for next year, but seldom if ever has a vision or plan as to the development of the company over the next five to ten years. Most of the management theories of recent years, and just about all the academic research and university degrees, have focused on the analysis of business to the extent that I heard the MBA (Master of Business Administration) degree at one famous US business school described as the 'Master of Business Analysis'.

Business is about making money: understanding the marketplace, developing economic (business or marketing) models, marketing, process planning, allocating resources and so on are all *parts* of the process – but managing the people, delivering the minimum performance standards, and owning and achieving the objectives are the real keys to success, because without these nothing gets done at all and all the marvellous modelling and analysis is a wasted activity.

I am not against modelling. I am, however, firmly of the belief that it is only a small part of the process. And, despite what people with MBAs, academics and many others may say, it is managing **people** and providing leadership in all its forms that will make or break a company. Your job, through the CQI Management Action Plan, is to change a set of attitudes so that managing and leading people becomes central to the way the company is run.

How can this be done? To answer this we first need to go back and look at the types of leadership that need to be displayed by managers in various situations.

In essence there are four distinct types of work activity: labouring, craft, technical, and knowledge.

Labouring

Physical activity applied to the process of making and moving things encompasses a great deal of the work done in any company. The work carried out may well require a considerable amount of training, but the necessary skills are physically repetitious and there is very little need for any knowledge beyond the basic schooling most people receive. Jobs in this category include secretarial and routine administrative work, most production and operations activities, and almost all the jobs involved in the storage, distribution, and display of goods.

Contrary to much of the current management thinking, the type of leadership or management required for this type of work activity is that of command and control. The goals set are immediate and orders are given (albeit dressed up as polite requests). The management style is directive, everyone knows what they have to do and they get on with it. Should any 'labourer' fail to do their job then the sanctions against them tend to be immediate, ranging from a dressing down to being fired. Once instructions have been given, most labouring activity does not require constant supervision and it would benefit both workers and managers if, having set the work in motion, managers trusted their people to do the job properly.

Craft

Once some acquired theoretical knowledge is required to make and move things, we are into the realm of the craft worker. Although their skills are learned by the repetitious performance of the task, they also apply theoretical knowledge that is taught to them by a more experienced craftsman or craft master. Into this category we can place the skilled machinist, senior secretaries and personal assistants, most of the marketing people, many types of sales activity, almost all the back office/administrative people who are not 'labourers' and so on. Again, since the work is to do with making and moving things, the leadership and management style that needs to be used is a more participatory version of command and control. Instead of issuing instructions, the manager should consult craft workers and then issue instructions.

In reality, when dealing with labouring and craft activity the leadership and management function is that of an overseer; the level of actual management and leadership skill required is not high and can, indeed should, be carried out by a senior member of the work group itself. This form of empowerment is both motivational and cost-effective.

Technical

As soon as advanced theoretical knowledge is applied to labouring or craft activity with the purpose of improving performance and productivity we move into the world of the technical worker. Historically, technical work emerged during the Agricultural and Industrial Revolutions of the last two hundred years – with most coming into existence within the last century. The knowledge to do technical work is taught (rather than learned from experience) and is applied so as to improve the way things are done and thus increase productivity.

Into this category of work we need to place all those in the company who are involved in the strategic and tactical planning of the business. Many will be in senior positions within each department and their job is to provide input into how the company can be advanced towards it goals. Management is itself a technical activity, but only where that activity is aimed at improving performance and not just to overseeing the work of others. Technical workers need to be managed differently from labourers and craft workers. They require goals: they need to be recognised for their performance in a clearer manner, they need coaching to a greater extent, they need to be motivated – they need *performance management*. These people will accept individual goals, whereas labourers and craft workers are more responsive to team goals. The best management practices I outlined earlier are most suited to the management of technical workers.

Knowledge

The knowledge workers are applying advanced theoretical knowledge to technical work with the aim of discovering solutions to problems we do not yet know about and thus developing ways of vastly improving the productivity of the technical worker. The senior strategic thinkers and planners, the marketing people who dream up new and exciting advanced products, the computer programmer who develops a better and faster operating system: in fact, all the visionaries in the company are knowledge workers and they need to be managed differently from technical workers. They respond to missions rather than goals, monitoring their performance becomes difficult since we will frequently not know exactly where they are going, and they will almost always know more about what they are doing than their managers do. Managers of knowledge workers need to protect their people rather than lead them and the manager himself or herself will merely be one of the team rather than the boss.

Given that different activities fall into different work categories, it is obvious that members of the workforce could well be carrying out different types of work as a normal part of their jobs and will, therefore, need different types of management leadership depending on what they are doing. Consequently, managers will need to be able to apply different management styles in different circumstances. But, as I am sure you recognise, many managers really only have one management style and cannot apply any others: these so-called managers are restricted in their value to the company *as managers* and, as part of the action plan, you should consider carefully the idea of moving them to a position in which their management style will be put to good use.

The main thrust of your action plan needs to focus on getting the right type of manager in the right management job so that they are in the most productive position according to their competencies and skills. This approach will, I can almost guarantee, result in a significant number of managers feeling unable to continue with the company as they will feel demoted or downgraded. But, and this should be stressed, the purpose of management is to use resources correctly, to manage appropriately and obtain the best possible performance from the team. If a manager is failing in any of these areas then he is failing as a manager and should be retrained or, if that does not work, reassigned. I realise that almost amounts to saying that some managers should be got rid of; but when a company is failing to produce the maximum performance of which it is capable it is the management who are responsible and, if 'down-sizing' is unavoidable, it is those in management positions who should be weeded out.

Before rushing to get rid of all your managers, however, you should first develop a comprehensive management training course that establishes the skills and techniques which should be used in the management of your people. This training programme, which would probably run for one week, should look at the various types of management style needed with the different types of work and the managers should be trained accordingly. It will also require you to use psychometric testing to determine the strengths and weaknesses of each manager – in terms of their style – so that weaknesses can be trained up and the strengths exploited through reassignments to more appropriate positions. In *Managing for Performance* I go into this in some detail and give examples of the tests that could be used to help you assess people – however, I strongly suggest you should also use an outside specialist consultancy (such as the authors of those tests) to assist you if you are planning to make reassignments.

All in all, your action plans need to result in a management approach that establishes full responsibility and accountability at all levels and rigorously but fairly applies sanctions for failure. If your company is to become a true CQI business then we need to make radical improvements in the way we manage people, we need to manage differently, and we need people to accept the responsibility for their goals and be accountable for the results of their success or failure.

SUMMARY POINTS

- Preparing the CQI action plans is a major part of the CQI process and requires additional resources in the form of work groups to focus on each plan.
- In preparing the plans you need to ask four interdependent questions:
  - What are we doing now?
  - Is it necessary?
  - How can we do things better?
  - How can we do things differently?
- The principal areas of focus are divided into three groups – 'soft' or behavioural (Human Resources, Training, Management), Marketing (Communications, Marketing, Sales) and Operational (Support Administration, Business Processes, Production and Operations).
- In all areas you need to establish the key benchmarking standards and the best practices you wish to have in the company.
- The Human Resources action plan should seek to estab-

lish that the following standards and practices are in position and fully implemented:
- full descriptions for all job functions
- career paths for all functions
- training prerequisites and commitments for all functions
- comprehensive HR development processes
- comprehensive objective (goal) setting processes
- performance review processes
- performance appraisal processes.

All of these should be clearly linked to the overall business goals of the company.

- The training action plan must be firmly linked to and derived from the overall HR strategy and it is necessary to complete the latter before trying to develop the former.

- Training is broken down into:
  - Knowledge
  - Skills
  - Coaching.

- Knowledge training focuses on the knowledge required to fulfil the job function up to and beyond the minimum performance standards for the job. Knowledge training should be explicitly required for each stage of development, and promotion to the next stage or grade should be dependent on its completion. There will be an ongoing necessity for knowledge training which is best provided internally.

- Skills training aims to provide the correct skills necessary for the job. It is generally a one-off activity with refresher courses and can often be most cost-effectively provided by outside training consultants.

- Coaching is a management activity to enhance and re-establish previously trained skills through on-the-job coaching.

- A survey carried out amongst leading companies shows that coaching is seen as the most important management activity if the company is to continuously improve its performance and succeed in reaching its goals.

- Coaching is a continuous activity and a formal programme should be established to ensure that every manager coaches all their team members on a regular basis of at least one day every quarter. The programme should include feedback from the team members on the coaching received so as to ensure that the activity is taking place as prescribed.

- The management action plan must focus on establishing the right managers and supervisors in the right positions.

- Managers must be made accountable and responsible for the achievement of the objectives of their teams and sanctions against failure should be in place and used in a fair but firm manner.

- Every type of work activity fits into one of four categories each needing its own type of management and leaderships:
  - labouring – directive command and control
  - craft work – participative command and control
  - technical work – goal based performance management
  - knowledge work – protective management through leadership.

- Managers should be trained and assessed on a regular basis to ensure that they have the skills and competencies necessary for the job they are doing – if they do not, then they should be reassigned to where their style will produce the best results. This may result in apparent down-grading (without loss of pay) and may lead to some managers preferring to leave the company – this is acceptable since it means that you are getting rid of someone who is not producing the performance the company requires.

- **Management accountability must become a reality, not just given lip-service, if your company is to achieve CQI.**

8

More Action Plans

Internal communications • marketing • sales, customer interface • support administration • business processes • production, operations

IN CHAPTER SEVEN we looked at the objectives to be achieved in the action plans for the 'soft' or behavioural areas. In this chapter we will be looking at the other areas of

- Internal Communications
- Marketing
- Sales
- Support Administration
- Business Processes
- Production and Operations.

There are, however, a number of things we need to remember if the CQI programme is to be effective.

(i) It is people, not processes, that are important. Unless we manage our people better and get them to deliver the maximum performance, we are misusing a costly resource and, no matter how much we change the processes, we will not achieve continuous incremental improvements.

(ii) There is no point in changing a process until we have changed the way we manage the people and have maximised their performance.

This point was brought out at a recent seminar. The delegates were examining the need for increased human resources for a project – the team leader wanted another person on the team and the seminar leader was asking why. In the course of a capacity exercise the delegates decided that, if each person was fully productive for 80% of the time (a not unreasonable assumption) then five people would be needed for the task – i.e., 4 man-days. In fact there were five people on the team, but the delegates calculated that none of the team was working at more than 60% capacity; they were thus delivering 3 man-days and so failing to achieve their goal. Obviously, if a sixth person was introduced and worked at 60% capacity (a likely situation given that the others were only working at that level), then the team would deliver 3.6 man-days; they would still fail to deliver their goal, but the overheads in staff cost would have gone up by 20%.

(iii) We should only change a process if we are seeking 'orders of magnitude' change in the productivity of that process – i.e. if we are seeking a 100% plus change in the effectiveness of the process.

(iv) Almost all processes can be improved by reducing wastage – in terms of both energy and materials. The only things that should be involved in the process are those things that **are really necessary**: anything else is a misuse of resources.

The key, therefore, is to deliver improved performance by managing our people better **before** setting out to change the way things are being done.

INTERNAL COMMUNICATIONS

The majority of staff surveys show that communications within companies are, in general, poor and this problem needs to be addressed as soon as possible. The communications strategy you have developed for the CQI programme is a good basis for an overall communications programme within the company although the 'branding' and 'image' for the general communications material will have to be rethought to reflect the different purpose.

You will need to review the processes used for internal communications: undoubtedly there will be a strong use of memoranda; if the company has widely spread locations then you may be using e-mail; and, of course, there are the telephone and fax machines. In addition, you should review the use of team meetings, departmental meetings, and management briefings at which policies and plans are explained to the staff. But most importantly, you should review what is being communicated and the way it is communicated – a somewhat more sensitive subject.

First, let us look at methods.

Perhaps the most important method that you should adopt is a regular staff newsletter or newspaper. This should cover everything from corporate policy to staff events, from company business to private news, and should always have pictures and text of any event at which senior management meet and talk to junior staff, especially if awards are being given. We all like to see our name in print and we all want recognition for good performance: if this is combined with a photograph of us receiving an award or just a picture of us or our team because we've done something newsworthy, then the boost to motivation is out of all proportion to the event itself.

But you must get the balance right within the paper and it must appear on a regular basis. One international bank puts out four to six issues each year of an A3 (tabloid size) newspaper with eight pages of news and photographs. This company newspaper is edited within the marketing department and produced by professional printers since the run is around 10,000 copies – one for every member of staff, distributed through the internal mail. The editorial team ensure that the balance is 50% reporting staff events and initiatives, 20% highlighting forthcoming events and 30% reporting management issues and policy. In terms of space, 40% is taken up by photographs, and the whole publication is produced in colour. In addition to this general newspaper, which is produced in English, the company also puts out quarterly business specific newspapers that adopt a similar format and editorial policy but with the additional requirement that 75% of the items must be about the individual business and only 25% about the company in general. These papers are produced by the marketing teams in the individual businesses who generally use desk top publishing and publish in the local language in black and white. Everyone in the individual business receives a copy and additional copies are circulated to other businesses for information.

It is interesting to note that before the introduction of the newspapers the company scored nearly 40% negative for internal communications and this dropped dramatically to only 12% after the newspapers were produced.

This particular company always used presentations as a method of getting information to the staff, but changed the approach slightly and introduced a more interactive process so that members of staff could question the speakers (normally senior managers) during the presentation. And to help make sure the information was acted upon a summary handout was introduced for all major presentations. To make this happen in a cost-effective way, every business installed a computer-based presentation programme which not only printed the overhead slides (or produced the 35mm photographic slides) but could be used to produce a summary of all the slides used. There was a 'house style' for each topic: senior management policy material had one style, sales material had another, and so on.

Other areas to look at include the use of 'news flashes' – single-sheet information briefings sent out via fax or the internal e-mail system. The recipients were asked to circulate them to all members of their team. These briefings were used to convey urgent or important information or to reinforce items in the company newspapers.

Some larger companies try to establish a 'house style' for the writing, but this is both difficult and expensive and tends to rob a message of its individual flavour. My recommendation is merely to circulate a manual to everyone outlining the key elements of good written communications and to encourage contributors to think before they write.

It is worthwhile stressing that information flows have to go up and down the company and that any blockages need to be cleared. Blockages generally occur (a) when too much information is circulated, and (b) if managers withhold information as a demonstration of their power. This last point can be true of national cultures as well as corporate ones.

Too much information is as dangerous as too little – each of us needs to have access to all the information we require to do our job and it is the responsibility of our managers to ensure that the information reaches us. However, we do not need more than that and other information of a more general nature should be made available through the newspaper or briefings. One company had a policy of providing a detailed breakdown of all the financial results for the whole company each month to every manager, only to find, when the communications team investigated, that a full 83% of the managers never looked at any of the results except those that directly related to their areas of responsibility – hardly a surprise. When revised MIS data was supplied covering their individual departments or teams, with a summary of the key financial ratios of the company, then usage went up dramatically and the consumption of paper for the reports was halved. Admittedly, the time taken to produce the reports increased, but this was balanced by improved productivity and performance by the MIS team.

If the blockages are caused by managers withholding information deliberately then this is a very serious problem and they will need to be trained and coached to change their behaviour. If that is not successful then serious consideration should be given to reassigning individuals or even getting rid of them. Harsh? Not really: such managers tend to hinder the productivity of their people rather than improve it and frequently have other bad management behaviours and practices as well.

But what should be communicated? This is a sensitive subject and one that raises considerable passions amongst management who often feel that some information should not be available to junior levels. Meanwhile, those at junior levels believe that unless there is full disclosure something sinister is being hidden by the management. This problem is more one of trust than of communication – when trust exists between the various levels within the company the problem tends to disappear.

If we go back to the principle that everybody needs certain information to do their job, we can see that any withholding of this information is counterproductive. But who decides what is needed? The answer is that all managers or team leaders should understand the jobs being done by those working for them and should, therefore, know what information is or is not needed to fulfil those roles. If there is a gap in this knowledge then tracking requests for information should reveal what information is being used by an individual in their work.

Having established what information is required, any other information is a distraction. However, people need to feel pride in their company, they need to feel part of the team, and this comes from knowing about the business, how it's doing, who is doing what, which people have joined the company, moved job or departed. This sort of information should be available to everyone through the staff newspaper, and especially information about the company itself – *staff should never find out about the company's problems from the national or international news media*. If they do, then it shows that there is little trust between the management and the staff and it puts the staff at a major disadvantage when dealing with their customers.

All this presents companies with a problem in deciding what information to make available and what information to withhold. In general, they tend to err on the side of openness and make information available unless there is a clear reason for withholding it. My approach, on the other hand, is to make available the information people need for their jobs, to publish company information that will help establish loyalty, pride and team spirit, and to withhold information that is commercially sensitive or confidential. Anybody requesting access to restricted information must, therefore, demonstrate a clear 'need to know' based on the work they are doing.

In terms of your action plan, you may well find that you can review the amount of information being circulated and reduce it dramatically. This not only lightens the burden on the staff but also reduces costs. If anyone misses the information or finds they need it for their job, then they can request it on an *ad hoc* basis. Not circulating every piece of information is not the same as restricting access to it – the information is available to those who find they need it.

One staff survey showed that a common complaint was that too many reports had to be prepared. A supplementary question revealed that they also felt that too many reports had to be read. The company reviewed the reports being prepared and cancelled around 80% of them – almost nobody commented on their disappearance or the reduction of information flow. I believe that what had happened in this case was that reports called for in a specific circumstance had not been cancelled when no longer required. A similar review within your company may well free a huge number of man-days.

MARKETING

Unless you are a marketing person yourself, preparing an action plan on marketing is likely to be beyond you. However, there are a number of things that, as the CQI leader, you should address.

Firstly, you will need to work closely with the marketing team to analyse the results of the external survey that should have been carried out. This survey will tell you how the customers perceive the company and the goods and services it offers. What we are looking for here is: are we providing the goods and services that the customer wants, in a way that they want, at a price they are prepared to pay and on which we make an adequate profit margin? In other words, is our product range customer-friendly, customer-focused and profitable? This has nothing whatsoever to do with the strategic decision as to what business we are in or what we are making and selling, or even the pricing strategy; it has everything to do with whether we are delivering our marketing strategy in a customer-oriented, quality service way that leaves our customers feeling good about doing business with us.

Marketing departments, if fulfilling the true marketing role of deciding what we will sell, at what price and in what market, will already be looking at new products and services, niche markets, leading-edge ideas, and will always be looking for a competitive edge. But experience shows that many marketing teams tend to assume that falling sales reflect a problem in the marketing mix rather than a failure to be customer-focused. The question that needs to be asked in these circumstances is: is the problem in the products or is it in the way we deliver them? Market research is vital here and will help identify the problem. If it is the products, then the marketing team needs to rethink. If it is a lack of customer focus, then it is a quality issue and by changing the delivery to one that operates in a problem-free, competent and timely manner and provides the customer with a satisfying quality purchasing experience, we can often reverse a declining situation and provide a boost to profits. After all, most companies operate in markets where even the most advanced products become generic very quickly and it is often the quality of the purchasing experience that provides the competitive advantage.

You have to remember, when doing the marketing action plan, that the customer does not perceive the company as a series of departments but as one integrated whole and his or her purchasing experience is determined not only by product features and price but also by the way it is sold, the way it is packaged, the way the sales administration affects them, the way problems, enquiries and complaints are handled, the way servicing is carried out, and a myriad other factors. It is important, therefore, to start the marketing action plan early so that any problems identified here can be linked to the action plans relating to other processes or departments so that the company is able to present a united face to the customer.

SALES AND CUSTOMER INTERFACE

This action plan is not only about the front line sales force but also about any part of the company that comes into direct contact with the customer for any reason. You should look at the people providing the after-sales service, customer enquiries, invoicing, customer complaints and so on.

The customer interface is the point at which the customer has a direct relationship with the company and it is the one area in which the contact will drive the entire perception the customer has of the business. If any of your colleagues have any doubt whatsoever about this then ask them to think back to the last time they had any contact with, say, their telephone company or their bank.

Let me illustrate: my wife has recently started her own business which she operates from home. Her contact with the client is frequently by telephone and I applied to have a telephone line installed for her own use. To do this we contacted the local telephone company, Belgacom (we live in Belgium), which has a very bad reputation as far as its customers are concerned. I had a different experience. I visited their 'phone shop', conveniently located in the entrance of a large local supermarket, and spoke to a very pleasant young lady who immediately found the necessary form and helped me complete it. The level of professional competence demonstrated was high – not only did she get the form completed, she also asked whether I needed an additional telephone unit (I didn't), how the number was to be entered in the directory, who was to be billed for the service, whether I was registered for VAT, and crucially, when I would be at home for the work to be done. It being a Friday, we agreed on the following Tuesday between 1200 and 1400 hours.

At 1200 on Tuesday the engineer arrived, fixed the line, went to the exchange to switch on the service, returned to ensure it was working, helped me connect the telephone unit, and was gone by 1315. The bill for the work arrived a few days later. My entire contact had been in English (not an official language in Belgium) and we were extremely pleased with the service rendered. I doubt whether we would have got the same level of professionalism, courtesy, speed, efficiency, and multi-linguality from BT or any of the other privatised European telecom companies – imagine the sales staff and engineers in BT carrying out their work in French! Belgacom's reputation is excellent as far as I am concerned.

But it could so easily have been a different story. The saleswoman could have just given me the form and let me struggle with it. She could have said that the engineer would turn up sometime on Tuesday. The engineer could have failed to arrive (ever heard that one before?) The whole affair could have been a customer's nightmare.

Or try a British high street bank. A typical story may go something like this: a customer who wants to buy some foreign currency has to wait in a queue for 10 minutes before reaching the counter. The teller carries on with whatever he was doing and then leaves his place to speak to someone elsewhere. Eventually he returns and condescends to speak to the customer who has now been waiting a total of 20 minutes. She states her request clearly and hands over her credit card – she wants £100 in Belgian francs. The teller completes three separate forms and a credit card form and hands most of them to the customer to sign, then leaves taking the credit card to somewhere in the back, returning five minutes later with the authorised charge. He then charges the customer £3 for the use of the credit card and starts to look for the Belgian francs. Eventually the customer is given her money, all the papers, and the credit card and, before she can put them safely away, the teller calls out loudly 'Next!' The customer has been in the bank for nearly 30 minutes for a transaction that should have taken less than 10 minutes. These events actually took place recently when a member of my family came to visit – all I can say is that the next time she will wait until she arrives in Belgium, where she will put her UK credit card in a local ATM (the ubiquitous 'hole in the wall'), receive her Belgian francs in approximately 2 minutes, and be charged the direct equivalent in sterling.

I am sure you and your colleagues can come up with hundreds of stories which demonstrate a good or bad 'purchasing experience' or 'customer-company exchange'. But it is not other companies we are concerned about – it is your company; and the market research and the customer satisfaction survey you will have run will tell you into which category – good or bad – you fall when dealing with customers.

The action plan you develop for the customer interface people should focus on the delivery of a service that is

- customer-friendly
- competent

- efficient
- problem free
- timely.

This is especially important when the customer has an enquiry or a complaint and this requires the person *who first deals with the customer*

- to **own** the problem or situation.

Here it is important for the staff member to recognise that the customer is not complaining about them personally but is concerned with something the company has done. What you must ensure is that your people deal with that situation without passing the customer from pillar to post, from one person to another. The first person the customer talks to should take down all the details, tell the customer what steps they will be taking to deal with the situation, explain it may take some time to sort out and arrange a suitable time to call the customer back, then go away and deal with the situation, and – whether the situation is resolved or not – **call the customer back at the agreed time**.

Your customers should always end their contact with your company feeling good and this requires that you deal with them in a way that reflects the importance of the position a customer plays in the day-to-day business of the company. Remember, the customer is not an interruption of our work, but the reason for it.

Obviously, training plays a major role in addressing service issues but having a suitable 'problems' policy in position is also vital. Everyone in the company must understand that the customer is not always right but is always the customer and is the most important person as far as the company is concerned. **This means that sorting out customer problems, complaints, or enquiries is a priority and not something to be done later**. Once this is firmly established it should be possible for anyone dealing with a customer to get the help and assistance they require as soon as they require it.

Equally obvious is that it would be far better if the problems did not occur and that your people are trained to such a degree that they can sort out any issue without having to resort to asking for help. This is where the 'mantra' of the quality movement means so much:

Do it now and do it right first time.

And this comes down to training and empowering the staff to do the right thing for the customer.

Your action plan should, therefore, focus on training. Training in the skills needed to deal with customers, training in how to deal with problems, and training in how to take action when required. It should also focus on giving the staff the authority to take action – something that many managers are reluctant to do. They see it as impinging on their status and yet, at the same time, they do not want to be constantly bothered with minor questions.

Empowerment of the staff is a key issue for a CQI company and it should be pursued with vigour through your action plan, but empowerment can only take place if the training has been given. And that training must reflect the importance of the customer interface staff in terms of the customers' perception of the company.

You can check how well you are doing in addressing these issues by running service quality records. For example: to see whether our customers are receiving the sort of service they want you could carry out satisfaction surveys on a contact basis – put a member of staff somewhere where they can contact the customer as soon as he or she has finished their business with the company. The questions asked should focus on the experience they had – was it good, was the customer interface staff member competent, polite, efficient, friendly, helpful, etc., was the business carried out in a problem-free and timely manner, was the amount of time the customer had to wait acceptable?

If you want to check that you are delivering a problem-free service then record the incidence of problems or complaints and also how long each situation took to rectify, how many people were involved, and whether the customer interface people received a problem-free service from other departments.

This sort of routine service quality tracking can be set up for any part of the business whether it deals directly with customers or not and the more frequently it is done the more effective it is at helping you improve the quality of service offered. But it should not become habit. Any service quality check should last only until some short-term goal for improvement has been achieved – if the goal is that 90% of all problems should be resolved within 24 hours (a typical service quality goal) then when a department has met this target for three months the tracking should be stopped and some other aspect looked at. The tracking of problems can and should be returned to later to ensure that the enhanced performance is being maintained.

Service quality tracking should not be confined to customer interface departments – every department is a 'customer' of other departments and this concept of internal customers is one that you will need to pursue with considerable vigour. One of the major failings of the quality movement is that most practitioners have failed to implement procedures for internal as well as external customers. Worse still, some have implemented it only for internal customers at the expense of external customers – this is something that a number of well-known businesses, ranging from public utilities to banks, have done in the UK and elsewhere.

Quality is important in every aspect of the business and treating internal customers and external customers in the same way ensures that quality service becomes a reality within the company. The checks and tracking that are implemented for external customers can be used, almost without modification, for internal customers as well and to make this a reality you will need to establish quality service goals for each department and monitor them thoroughly. The most effective quality companies make reporting on quality initiatives and quality goals mandatory for all levels within the business and some have even gone so far as to link bonuses and other benefits to the achievements of quality goals.

SUPPORT ADMINISTRATION

Beyond the implementation of service quality tracking, your action plans must also look at improved productivity and better utilisation of resources within the support and administrative functions.

Care must be taken to ensure that those preparing the action plans do not focus entirely on reducing wastage as this can, and frequently does, result in restrictive rules being established which, although aimed at reducing costs, reduces rather than enhances the empowerment of the staff. No one can condone what is often called staff pilfering – making private calls on company phones, writing personal letters during working hours and on company paper and sending them through the mail room, taking stationery, pencils, etc. from the stock cupboard, and the thousand and one other ways that employees use the company's resources for their own purposes. But empowering staff, making them responsible and accountable for what happens in their area and department, and genuinely giving them a stake in the business through bonuses and share options, leads to vastly reduced wastage since everyone has an interest in maximising the profits and efficiency of the company.

Reduction of waste must come as a genuine by-product of a better managed and empowered workforce rather than by enforced application of restrictive rules. For this reason the action plans should focus on removing barriers to the efficient running of a section, team, or department

– what one of my colleagues refers to as eliminating the 'Mickey Mouse' in business.

Some classic examples of 'Mickey Mouse' practices we have come across include:

- secretarial and typing staff who were instructed to take coffee and tea breaks at specified hours and for a specified period of time – when the restriction was eliminated overall efficiency and productivity rose;
- a company that effectively closed down in the middle of the day because everyone had to take lunch at a specific time – when lunch breaks were made flexible there were always people in the office and a number of profitable deals were done;
- a major exporting business whose switchboard closed down at 5 o'clock just when the US market was most busy – when everyone was able to make domestic and international calls without having to go through the switchboard the US exports soared without the telephone bills going up by much;
- the manager of a major administrative office who was not able to obtain stationery for his staff except through central purchasing who only worked half-days – the manager was given stock to hold within his department so that ordering only had to be done on a weekly or monthly basis; and there are many more.

Eliminating petty restrictions, trusting the staff, and empowering everyone within reasonable limits is the key to vastly enhanced productivity, improved efficiency, decreased costs and increased profits. And this should be the aim of the action plans – but you will need to empower each department to prepare their own action plans and make them responsible for achieving certain CQI goals.

Business Processes

Business process re-engineering is one of the great management theories of the 1990s. Since the phrase was first invented by Michael Hammer and James Champy in their 1993 book *Re-engineering the Corporation* many companies have embarked on massive process re-engineering – often, I regret to say, without a full understanding of the implications of what they were undertaking. In late 1994 and early 1995 Hammer and Champy were warning that around 80% of re-engineering projects were failing – mainly because they had been started without the real long-term commitment of senior management and for the wrong reasons.

Hammer and Champy defined re-engineering as 'the fundamental rethinking and radical redesign of business processes to achieve dramatic improvements in critical, contemporary measures of performance, such as cost, quality, service, and speed'. They then went on to say that the statement contains four key words: fundamental, radical, dramatic, and processes.

Fundamental

In their opinion re-engineering takes nothing for granted – it ignores what is and concentrates on what should be. It is not concerned with improving a process, it is questioning whether that process really needs to exist at all.

Radical

They believe that re-engineering is about reinventing the business – it is not about business improvement, business enhancement or business modification. If all the company needs or wants to do is sharpen the edges and improve the way they do things, then re-engineering is not what they need.

Dramatic

'Re-engineering isn't about making marginal or incremental improvements but about achieving quantum leaps in performance.' Achieving 10% improvements can be made by incremental quality programmes (such as CQI) or by other more conventional methods – re-engineering should only be undertaken when the improvements required are in the order of magnitude of 100% or more.

Processes

This is a vital word but 'it is also the one that gives most corporate managers the greatest difficulty. Most business people are not "process-oriented"; they are focused on tasks, on jobs, on people, on structures, but not on processes.' They go on to define a process as 'a collection of activities that takes one or more kinds of input and creates an output that is of value to the customer'.

Re-engineering of business processes has little part to play within a CQI programme – only if the desired incremental improvement cannot be achieved in any other way should it be undertaken.

But that is not to say that your business processes cannot be improved. And, indeed, your action plans should focus on ensuring that all processes are made as efficient as possible. Initially, however, you will probably have to confine yourself to examining the functions and activities that make up the processes rather than trying to examine processes themselves. This, of course, has little to do with re-engineering but it is important to remember that re-engineering should only be attempted once you have obtained all the incremental improvements possible.

The first step, working with those people who are already doing the jobs and fulfilling the functions, is to record exactly what is involved – each step, each part of the paper flow, each movement of goods or whatever. This is a time-consuming activity and one that is best undertaken by a team composed of a CQI person, the job holder and their supervisor. These three can then write down what is done, what should be done, and what could be done. And at each step the team should ask: is this action or step necessary? Could we do it differently? Could we do it better? (This is not the same as the far more fundamental re-engineering question: is this process itself necessary?)

When process improvement is started it is quite common to find a range of things that could be done to improve the process immediately. Some of these improvements may affect other parts of the process and other people: if so, then they should be identified but *not implemented* until everyone affected by them has been consulted. For this reason it is wise to ensure that the CQI team have briefed each group of people as a whole on what is being undertaken.

The importance of this was brought home to me when I was working in a bank. When the people responsible for establishing deposits received an instruction from the customer, they would complete a short form in triplicate. The form had to go to a 'checker' who signed it, then to a second 'checker' who authorised the transaction after confirming the funds were available, and then was sent to the treasury department where the deal was booked. The form then came back to the original person who entered the transaction on the customer's account. The whole process could actually take up to three days. It did not take much investigation to see that the procedure could be computerised and that the checkers (who were there for security and audit reasons) could authorise directly on to the computer and the treasury could also access the deal via the computer. But then someone asked why we needed two checkers and that opened up a can of worms – the audit people insisted on a paper transaction slip for the file and security wanted a separate copy for their files. The attempt at improving the procedure nearly floundered at that point. It was rescued by someone getting security, audit, treasury, corporate management and the front line officers together and agreeing that something needed to be done. Each then went away and started an improvement process that was interlinked with the original plan. Eventually, fifteen action plans were implemented as a result of trying to improve just one process.

This sort of linkage is very common and can well prompt the suggestion that 'we should just re-engineer the whole process', but that would be false thinking. Here we have not one process but a series of processes that are interlinked and all have different reasons for their existence and different desired outcomes. In my illustration, the security department has internal security and protection against fraud as a desired outcome. Audit has adherence to corporate, national and international banking regulations as its desired outcome. Treasury has trading and the original person has satisfying a customer requirement as their desired outcomes. It would be a far better approach to seek incremental improvements in each process which may well lead to an overall improvement greater than the sum of the individual improvements. It is only when no further incremental improvements are possible that radical redesign should be considered.

Production and Operations

Production and operations are, of course, business processes in themselves but they have a number of additional factors that have to be taken into account when seeking continuous quality improvements. Not the least of these is that the direct output of most production and operations departments is the product or service that the customer comes into contact with when they chose to do business with the company on a day-to-day basis. For example: a working system is what customers want when they use the telephone and the provision of that system or service is the outcome of the telephone operations department. Similarly, a motor vehicle is the outcome of the production department of a vehicle manufacturer and that is what the customer buys. Quality in the production and operations department has a direct impact on the customer – poor quality will be spotted immediately and almost certainly result in a decline in sales of the product.

The Japanese have really understood this. You only have to look at a Japanese car – even a bottom of the market vehicle – and it becomes apparent that all those things which European manufacturers euphemistically refer to as 'optional extras' have been fitted as standard; with a European car you will have to spend another 10–20% of the price to obtain the same specification. Alternatively, consider the likelihood of something being wrong with the car, something that needs to be rectified; the chances are that at the first service a European car will have a list of things to be fixed whereas the Japanese model just needs an oil change and a quick once-over. A third area is in the delivery times: I have just ordered a German-built four-wheel-drive vehicle and have been told that the delivery time is eight weeks – if I had ordered a similar Japanese vehicle it would have been available in under four weeks.

Just-in-time (JIT) is almost a credo in Japan. Suppliers and manufacturers work closely together to ensure that no one has to hold stocks of anything (thus reducing the costs of warehousing and enhancing cash flow) but, hand-in-hand with this, everyone ensures that what they deliver is manufactured precisely as required and that the defect-incidence is as close to zero as possible and certainly not known to the customer. If a defect occurs on a production line then the line is stopped until the defect is corrected so that all the goods leaving the factory contain zero defects. Of course, it does not work all the time – after all, humans are involved – but it does work 98% of the time in those businesses that have adopted the process fully and have persuaded their suppliers of the same need.

Linking quality to bonuses is the way that many companies make sure that their people become quality conscious – and they do it on a team or department basis so that it is up to the team to ensure that quality standards are adhered to if they wish to receive their bonuses. One bank insists that each member of staff has their primary bank account with the company – staff members soon register if there are poor quality standards in the operations department as it shows up on their accounts, and knowing you'll suffer along with the other customers is a good way of ensuring that quality service is offered to everyone.

To implement quality standards in production and operations departments often requires a fundamental change in the thinking of those people involved and establishing a '**Do it right first time and eliminate waste**' approach. That is not going to be brought in overnight. The setting up of quality teams, quality circles, quality standard tracking (such as defects or problem incidence), are all part of the process and you should involve your production and operations people fully in the CQI programme – in fact, once quality is a way of life in the production and operations area then almost all other departments will naturally follow suit.

SUMMARY POINTS

- It is people not processes that are important – unless we manage our people better and get them to deliver the maximum performance, we are misusing a costly resource and, no matter how much we change the process, we will not achieve continuous incremental improvements – there is, therefore, no point in changing the process until we have changed the way we manage the people.
- We should only change a process if we are seeking 'orders of magnitude' change in the productivity of that process – i.e. if we are seeking more than 100% improvement in the effectiveness of the process.
- Almost all processes can be improved by reducing wastage – in terms of both energy and materials. The only things that should be involved in the process are those things that are really necessary – anything else is a misuse of resources.

- A major improvement in the quality and effectiveness of internal communications can be achieved by publishing a regular staff newsletter or newspaper. This should cover everything from corporate policy to staff events, from company business to private news, and should always have pictures and text of any event at which senior management are involved with junior staff.
- Presentations with slides are a very good method of informing staff of planned events or changes in policy. They are especially useful as a method of providing contact between senior management and junior staff but they must be made interactive – staff should be encouraged to ask questions and should be provided with a summary handout at the end.
- Single-page 'news flashes' sent by fax or e-mail are an effective way of conveying urgent and important information and can be used to support the staff newspaper. Care must be taken to ensure that such 'news flashes' are circulated to all members of staff.
- Information must reach those who **need** it for their jobs. This flow is both upwards and downwards and blockages can and do occur, generally when too much information is being circulated or when managers believe that controlling the flow of information gives them power.
- Regular reviews of the number, quality, and necessity of the reports prepared in the company should be a priority. The reports deemed necessary should be retained and made available (but not necessarily circulated) so that those who need them can have access to them. All other reports should be scrapped.
- The CQI programme's main input to the marketing action plan should be to help the marketing team understand and review the results of the external customer surveys. These surveys tell us whether our company,

with its products and services, is perceived as customer-focused and customer-friendly.

- There is a tendency for marketing departments to assume that falling sales reflect a problem in the marketing mix or the market positioning rather than a failure to be customer-focused. With products and services often becoming generic very quickly and with increased competition, it is the quality of the purchasing experience that provides the competitive advantage.
- It has to be remembered that the customer perceives the company as an integrated whole and not as a series of separate departments. To change the customer's perception of the company means improving the quality of the service provided by all the departments – it is important, therefore, to start with marketing as this is where you will find out what the customer thinks.
- The customer interface is the point at which the customer has a direct relationship (or contact) with the company and the way that interaction goes will drive the customer's entire perception of the company. It is important, therefore, that **all** customer interface staff should deliver a service that is customer-friendly, competent, efficient, problem-free and timely. Training to deliver this type of service is of critical importance for everyone who deals directly with the customer in any capacity from sales to administration, from marketing to handling customer complaints.
- The key to customer focus is to remember that the customer is not an interruption of our work but the reason for it; and, although not always right, the customer is always the customer and the most important person as far as the company is concerned. This means that dealing with the customer is a priority for everyone involved and is not something that can be put off until later.

- If your people are to 'do it now and do it right first time' as far as the customer is concerned then they will have to be empowered to do the right thing for the customer. Empowerment is a key issue for a CQI company and must be carefully but thoroughly introduced at all levels.
- The quality of service received by the internal customer is as important as that received by the external customer. Everyone in a company is an internal customer for the services and output of other departments and the same quality tracking procedures should be adopted to ensure that the internal customers receive a quality service.
- The most effective quality companies make reporting on quality issues and quality goals mandatory for all levels within the business and many have even linked bonuses and other benefits to the achievement of quality goals.
- Improved productivity and improved efficiency in the utilisation of resources is a priority for all support and administrative functions, but care must be taken to avoid just focusing on reducing wastage as this can and frequently does lead to the imposition of restrictive rules.
- Empowerment of the staff, making them responsible and accountable for what happens in their area or department, and giving them a genuine stake in the business through bonuses and share options generally leads to vastly reduced wastage and improved efficiency since everyone has a personal financial interest in maximising the profits of the company.
- Process re-engineering should only be undertaken if orders of magnitude improvements need to be achieved – in all other cases obtaining incremental improvements in productivity and efficiency through CQI should be the aim. This is best done through close examination of

the functions and actions involved in each job function and then asking: is this step necessary? Could we do it differently? Could we do it better?

- The output of the production and operations department has a direct and immediate impact on the customer – they are the departments that make or supply the products and services – and quality in these areas is vital. Failure to obtain quality improvements here can lead to an overall failure of the CQI programme.
- Most of the work in the quality movement has focused on the production and operations departments in recognition of the key role they play and they should be your priority area. Quality improvements here lead naturally to quality improvements elsewhere in the company.

Part IV

Achieving Continuous Quality Improvement

9

Implementation and the CQI Cycle

Implementation • monitoring and feedback • reassessment and modification

WITH ALL your action plans prepared the whole CQI process will have gained momentum and you can move confidently into the implementation phase of the programme. And once implementation has begun you are into the Continuous Quality Improvement Cycle as illustrated on page 184.

IMPLEMENTATION

The first step in implementing the action plans is to encourage the individual departments to commit a small number of people to the CQI programme so that they take responsibility for the action plans.

The need for commitment from each department has been stressed and you will probably have co-opted some people already, but now each department will have to take responsibility for its part of the programme. This may well mean that the action plans, the benefits of CQI and the overall need for the programme will have to be sold within each department. And I mean sold! The commitment of the people involved is vital for the success of what you are

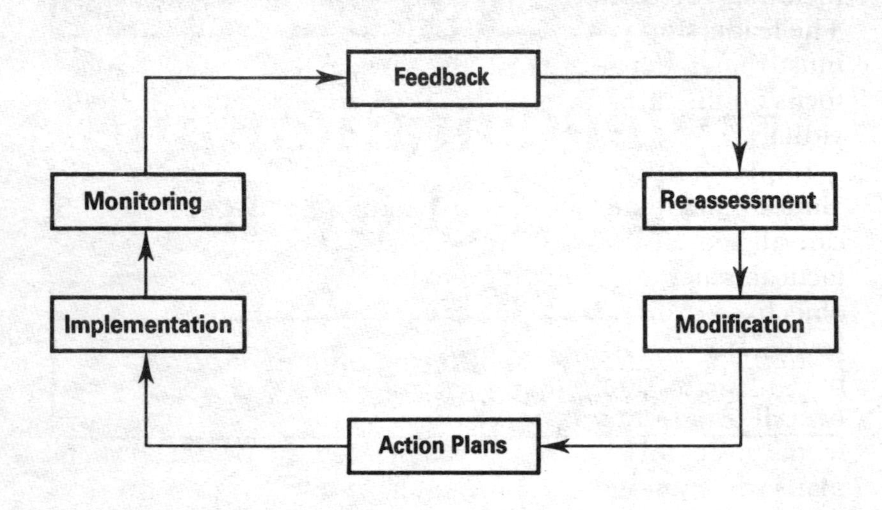

The Continuous Quality Improvement Cycle. This cycle shows the linkage between the action plans which you have now completed and the way that continuous quality improvements are obtained.

doing and that can only be achieved if they have a clear idea of the benefits involved. To achieve this you should revisit your original CQI communications strategy and refocus it at departmental level.

But selling the process is unlikely to be enough and a certain amount of enforcement may be necessary. This is achieved by involving the senior management, and especially the CQI 'godfather', and arranging for them to require clear reporting at all levels on implementation of the CQI action plans and performance towards the CQI goals. This reporting requirement must be a priority and not a gesture – implementation and progress towards the CQI goals should be linked to bonuses and other soft benefits such as additional holiday entitlements. Such a linkage will demonstrate the importance the company is placing on CQI.

Strong and clear leadership will also have to be demonstrated by you and all members of your CQI team including those newly recruited within the departments. The leadership you give will have to be very visible but you must remember to balance the three parts of leadership – focus on the task, focus on the team, and focus on the individuals.

Implementing well-thought-out plans and seeing the almost immediate results is an exhilarating experience and can all too easily lead you into involving yourself in the tactical issues at the expense of the larger, overall strategic objectives. You and your main CQI team must always review any CQI action in the light of both the tactical objectives of the action plan and the strategic objectives of the overall programme. On the other hand, the departmental teams need to focus solely on the tactical issues of their plans while not ignoring the implications of their actions at company level. As the implementation proceeds, issues will arise that have an impact outside the department and the leadership by you and your team must be towards maximising the effect of good ideas and minimising the impact of actions that, while necessary, are disruptive to others. Your role of co-ordinator, peacemaker and task master requires you to manage upwards and downwards as well as 'managing' your colleagues.

'Managing upwards' means getting your superiors to do what is necessary for the programme to succeed. Obviously, you cannot give orders to those above you nor can you set them goals and establish penalties for failure: however, you can and must work closely with them and keep them committed to the process. How? Through good and frequent communications, informing them of the successes as they occur and advising them of problematical situations in which you need their assistance. As one senior manager said to me: 'Tell me about what has gone well so that I can see the benefits to the company and tell me when you need help and why you need it. In this way I can give you the back-up you require, when you require it, because I know the benefits it will bring.' Managing upwards is as much a skill as managing those below you and your direct colleagues – each requires a different approach but all require you to 'sell' what it is you want done. You will have been appointed to the CQI role because you can get the job done and because more senior managers are busy with their own responsibilities – they will appreciate being informed and be willing to help, providing they are not asked to spend too much time away from their main job.

Managing your colleagues is, in many ways, a very similar exercise. Again, you cannot give them orders but you do have the advantage that you can make use of the senior management to help get things done. The problem is that your colleagues may believe they have things to lose through participating in the programme: they may feel their job is under threat or that their authority is being eroded through the empowerment of their staff. To overcome this you may well have to spend a considerable amount of time selling the benefits of CQI and the action plans and objectives for their department.

A colleague whom I will call Sue felt so threatened by the programme that she passively resisted it. She was able to get away with not carrying out any of the actions called for until one day it dawned on me what she was doing. I travelled to her office (which was located in a different country) and in a meeting with Sue and her boss I innocently enquired what was delaying the implementation of the action plans. Her answer told me far more about her insecurities than it did about problems with the plan. I was careful not to attack her and it soon became clear that she thought she should have had the job that I was doing and, if she could not, she was prepared to resist what I was trying to do.

I was stumped as to the way forward, especially since her boss was equally resistant, and I had to ask for help from senior management. The business manager quietly let it be known that progress with action plans and CQI goals was to be the first item on the agenda at the next management meeting and that he expected each manager to report success. At the meeting every department and team reported major advances except for the team run by Sue. Being an intelligent person, she got the message: however, she still did as little as possible and soon afterwards she was reassigned to another part of the company. A clear message was delivered: the business was serious about CQI.

While having to manage the management, you also have to support the CQI team and the change managers – those leading the CQI effort in each department – and this can also present unexpected problems.

The first problem is that most managers, in my experience, delegate the CQI role to the most junior person in the department. More than once I have found that the CQI team leader is a secretary, an operations clerk, and even on one occasion the odd job man. By the nature of their functions, these people tend to be low on the authority scale and have a great deal of difficulty in getting others to do what they want. Usually they also lack managerial or leadership qualities, although I know of a number of exceptions. Your job is to supply inspirational leadership to such people and to lend them the weight of your own authority until they have established a method of operation that will ensure that things get done. It may also be necessary for you to arrange for controls to be put in place so that the department head or team leader is fully accountable for the CQI results – in this way he or she is more likely to be supportive.

The second problem I have found is that although the CQI team may be enthusiastic about the programme, very few others share that enthusiasm and you will need to spend a good deal of time helping your team to sell the ideas into their departments and teams. In many ways this is a repetition of the selling you will have had to do at higher levels. It could be a very good idea to involve senior management at this critical stage so that everyone is aware that the company is serious about gaining continuous quality improvements. I found that I needed to spend a good deal of time with the managers of the teams and departments and then make presentations to their staff –

all the time ensuring that the designated CQI person was with me and was seen as enjoying my full confidence. This often meant that I had to take time to train them in what to do and how to do it, and in this I was supported by my own CQI team who made frequent and regular visits to all departments and teams.

Maintaining enthusiasm and building people's self-confidence is another major part of your role. At the beginning of the programme most CQI implementers run into a great deal of resistance as they try to get the action plans going and this tends to sap their confidence – you have to step in and act as a shoulder to cry on, a cheerleader, a coach, a father or mother figure, a supporter and a leader.

Generally speaking, though, at this stage in the programme, and if you've had a good and active communications process in place, people will be well aware of the CQI Action Plans and expecting them to be implemented. Clearly, therefore, the communications process is critical to the success of the programme and the more things are pre-sold to the staff the easier it is to get the plans implemented. Having said that, I must stress again that it is the managers who are the key; if they are supportive the action plans will be implemented easily, if they are neutral it is slightly harder work, but if they are against the programme then you will need all your persuasive skills and authority to get things done.

Given that success is built on an experience of success it is vital that every advance made must be recognised and publicised. This is one of your most important roles. Bland reporting of CQI results is definitely not enough – a song and dance must be made of each and every milestone reached in each and every plan. Get the senior managers down to the teams and departments to be told of their successes. Get them to make awards and participate in team meetings. If a new idea arises that will lead to further incremental improvements, make sure everyone in the team or department knows about it – and then make sure everyone in the company knows about it. But most importantly, make sure that the person who has achieved the success or come up with the idea is recognised and, if appropriate, rewarded for their achievement. Get people's names into the company newspaper. Get the senior managers to make a personal acknowledgement either by turning up and thanking the person (the best way) or by sending them a personal message and copying it to that person's manager. All this cheerleading is vital to build the enthusiasm and commitment of everyone.

MONITORING AND FEEDBACK

There are two distinct levels of monitoring that need to be established and carried out. The first is against the milestones contained within each and every action plan, and the second is against the milestones and benchmarks established for the whole CQI programme.

When the action plans were prepared you will have ensured that each plan had an overall objective – something that had to be achieved. Such objectives are likely to take months to achieve, some may take years; but each objective can only be reached by achieving other more immediate goals or milestones and you should track the performance of the programme against these short-term goals.

Take, for example, the Human Resource management plan. This plan calls for every job function to have a full written job function description attached to it along with the associated training programmes. In a small company this may take a month or so of dedicated work, but in a large organisation of a few thousand people it could well take a dedicated team over a year to deal with. In the meantime, short-term goals could be put in place: all clerical functions to be documented within three months, all training programmes attached within six months, all production and operations functions to have descriptions within nine months and so on. As each of these goals is achieved the team knows it is getting closer and closer to the final objective of the plan – this, in itself, is highly motivational and it also allows you to know how the plan is going and whether it is on target.

Each department and each team will have its own action plan and each action plan must have its own milestones. As each milestone is reached the team should be congratulated and recognised and the milestone should be ticked off on the schedule and noted down on the master plan. As more and more milestones are passed the impact of CQI will begin to be felt.

The passing of action plan milestones also affects the main programme and each builds towards achieving a milestone within the overall CQI Master Plan. Once people start to see the milestones being ticked off on the Master Plan, then the momentum will build. To achieve this early on is important and you should ensure that there are a number of reasonably easily achieved milestones for the start of each plan and some straightforward milestones for the Master Plan so that progress can be seen to be made right from the beginning of the programme.

As well as monitoring against the milestones within the plan you will also need to monitor the changes against the benchmarks you established in the research phase of the programme. This means that you will need to repeat the surveys at periodic intervals as discussed earlier.

And I mean *repeat* the surveys. I do not mean run new surveys. If direct comparisons are to be made then the method of measurement must be the same and this means using the same surveys as were used to establish the benchmarks in the first place. The same questions analysed in the same dimensions, the same style and layout, the same question order must all be used – in this way a direct and accurate comparison can be achieved. After all, if you are measuring the altitude of an aeroplane so that the pilot can maintain a constant and safe height above the ground it would not be wise to suddenly change the altimeter reading from feet above sea level to metres below an orbiting satellite. Nor would it be wise to move the altimeter from in front of the pilot and place it in the ground control.

A word or two of warning: do not run surveys too close together or too far apart. My preferred approach is to run each survey at 18-month intervals but twelve to 24 months is acceptable. Any longer than that and your people will have forgotten the previous survey and its results and what you are trying to do will be lost – each new survey is to tell everyone how the business is progressing against the benchmarks. Too close together is likely to result in too small an incremental change having taken place, which can be demotivating.

As always with monitoring, you need to make sure that the results are available to everyone in an easily understood form. An approach I have found very successful is to monitor each action plan and carry out formal reviews of progress on a monthly basis and then indicate on the CQI Development Curve the result as shown below.

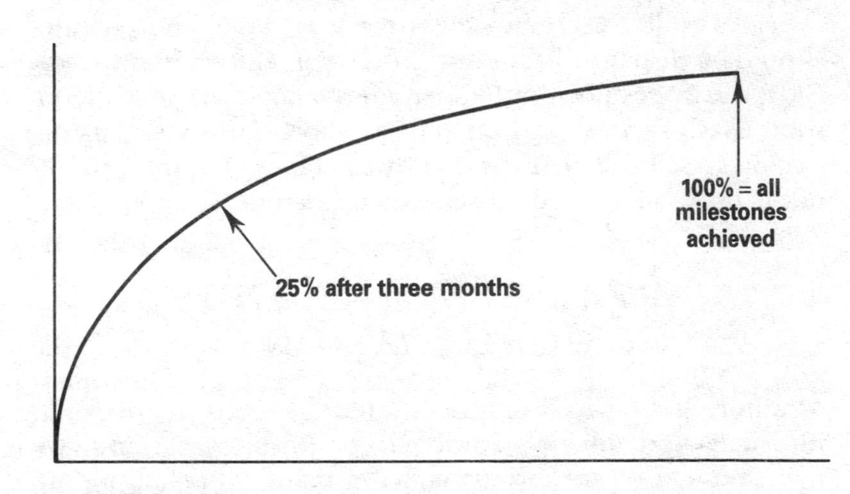

The CQI Development Curve showing that this action plan has achieved 25% of its milestones after three months.

Regular production of these graphs for each action plan followed by public distribution or display is the key to gaining commitment and motivation, the power of which should never be underestimated. Another useful chart is a matrix showing the milestones and a box into which the date of achievement can be inserted. This chart, when displayed with the Development Curve, gives everyone a clear idea of how far they have got and exactly what has been achieved.

Some companies prefer to circulate the results through 'news flashes' while others take the opportunity of having the results to hold a team meeting. Whatever the chosen method, you need to ensure that everyone in the team or department receives the results as soon as they are available. The results of the Master Plan can be treated slightly differently and made available through the newspaper (which goes to all staff members).

The benchmarking scores are, of course, company-wide results and these should be distributed through meetings or the newspaper. I prefer the former as a meeting allows the staff to ask questions about the surveys and about the overall results. As each new survey is done, the change should be clearly indicated – that is what matters in terms of CQI. Again, results should be made available to everyone as soon as is practical – whether they show improvements or declines against benchmarks they can be used to the benefit of the programme and must be seen as a powerful tool.

Reassessment and Modification

Monitoring of performance should be used to provide informal and formal feedback to the teams and to management on the progress being made (after all, we all like to see and experience our progress). But most importantly, the feedback should be used to modify and improve the action plans themselves.

No complex set of actions can be executed precisely as planned – unforeseen situations may block forward motion, a better route may be revealed, implementing one part of the plan may be so disruptive that an alternative has to be found, a superb idea may come to light as a result of early actions. All these things mean that the implementation does not go as planned and we have to reassess and modify what we are doing.

When a tactical action does not take place as it should, then we need to know (a) whether we have failed to take into account a significant factor – either known but ignored or unknown, and (b) what is the impact of the event on 1) the tactical action plan and 2) the strategic master plan.

The first step is to assess what happened – just because it was not what was planned does not mean it is a failure. I am sure you have, in the past, taken actions based on assumptions or incomplete information and the error becomes apparent very quickly but does not necessarily require you to stop or back-track; sometimes the new situation may be more to your advantage. But the golden rule is always to reassess the situation to decide whether what has happened is to the advantage of the tactical action plan **and** the strategic plan. And I do stress, we have to look at both plans – the situation may make implementing the tactical plan easier but may not achieve the strategic objectives and it is these that are of paramount importance in the CQI programme.

I have made the mistake of forgetting this golden rule. I was implementing an action plan within a business unit of one of my customers that required all members of the unit to be trained in a new operating procedure. It soon became apparent that many of the staff had already adopted a procedure that was very close to what was planned so that the training was apparently unnecessary. I took what looked like a good opportunity to complete the action plan early and failed to consider whether the procedure now operating was going to deliver the strategic objectives associated with the overall business process. Unfortunately for us all I had made the wrong decision and we ended up having to do the training anyway – in fact, it was more difficult to implement the new procedure since the staff had to unlearn the one they adopted. If I had taken the time to examine the procedure they were using in the light of the objectives of the modification of the business process, I would have seen that there were significant differences that were predominantly driven by the needs of the particular department.

But the result is not always negative. Sometimes situations arise which will improve the prospects of achieving the strategic objectives. You must take care to ensure that this is really the case but if it is so then you can modify the action plans accordingly. No tactical action plan is cast in stone – flexibility is everything, especially during the implementation phase, and reassessing and modifying the plans is an important tool that can help you reach your objectives sooner.

Reassessment and modification are also important as action plans achieve their tactical goals – which are themselves steps along the way to achieving the strategic objectives of the CQI programme itself. And it is your direct responsibility to ensure that any modifications to action plans – either during implementation or on completion – will result in an incremental improvement for the company. This means that you have to continually update the CQI Master Plan using input from each action plan and, as each tactical goal is achieved, establish a new goal.

It can be confusing to people to be faced with a situation which has no final, achievable objective – a situation in which continuous incremental improvements are the objective – and they may well find it difficult to adjust to the need for constant change. But that is what CQI is about. Although the business will eventually achieve the full implementation of what you have decided are the best practices, this is but a step on the CQI path, they are not an end in themselves. To maintain the momentum towards CQI you must always be looking beyond the best practices and the benchmarks and seeing where other incremental improvements can be achieved. New action plans must be prepared before existing ones are fully implemented and new strategic objectives agreed before the existing ones are reached – only in this way will you and your colleagues be able to take the 'step in faith' that will take you beyond best practices and beyond short-term benchmark achievements.

This movement beyond the current plan requires that all the activities involved are closely co-ordinated and that conflict between the objectives of the tactical plans and that of the strategic Master Plan is avoided. You will need to prioritise actions and tasks and you will need to be at least one step in front of everyone – you will need to deliver visionary but task-focused leadership and you must be prepared to adopt new best practices and new benchmarks. This may well result in you having to invent things since these new best practices and benchmarks may well not exist and other companies may well be looking to your business as a source of what is best in the quality field.

SUMMARY POINTS

- Continuous Quality Improvement comes as a result of the CQI cycle shown on page 184: Action Plans – Implementation – Monitoring – Feedback – Reassessment – Modification – Action Plans.
- Implementation of the action plans requires assistance in the form of dedicated staff within each team and department who will take responsibility for the programme.
- Action Plans need selling to the managers first. Once they are 'on side' you can sell them to the teams. This requires an active and forceful communications strategy – it also requires input from senior management who

should be encouraged to establish CQI goals as a reportable item for management and link them to bonuses and other benefits.

- Getting the action plans implemented means managing upwards (senior management) as well as managing your colleagues (who may well feel they have much to lose) and managing everyone involved even though they do not report to you – your role is that of co-ordinator, peacemaker, task master and cheerleader.
- Problems will occur, that is just a fact of life. You must move quickly and authoritatively to solve the problems and help the CQI teams to remain enthusiastic, committed and focused. You must also build and rebuild their self-confidence.
- The performance of the departments and teams in terms of the milestones and goals of the action plans should be monitored in a public but simple manner and the information made available to everyone. Success is built on the experience of success and people experience their success by seeing their performance in terms of their goals.
- The feedback from the plans should be used to reassess the plans themselves and to modify them if appropriate. Flexibility is vital but changes should only be made if they will lead to the more rapid achievement of the strategic objectives of the CQI programme.
- There is no end to a CQI programme and, as each action plan goal is achieved a new one should be developed so that further incremental improvements can be obtained.

10

Now That We've Got Here, Where Do We Go Next?

CONTINUOUS QUALITY Improvement is a journey that has no end although many people will feel, from time to time, that enough has been achieved and that more improvement is not worth the effort. But each incremental improvement is just another step along the path and how far you and your company choose to travel along the road is entirely up to you.

I have called CQI 'a strategy for survival' and, as products become generic faster each day and competition increases, it may become the only strategy for survival. In the past, bigger often meant better and economies of scale became the goal for many companies; today, however, we are faced with what John Naisbitt (author of *Megatrends* and *Reinventing the Corporation*) calls 'global paradox' – the bigger the world economy, the more powerful its smallest players. Companies all over the developed world are down-sizing, right-sizing, de-layering and generally becoming smaller as they seek to re-focus on their core business activities and obtain a competitive advantage. Many well-known international and multinational businesses are becoming niche players, no longer interested in being all things to all men but preferring to be the supplier of choice to a defined and carefully selected market segment. The products and services they supply are frequently the same as those of their competitors and, in quite a few cases, their price is higher and margins more profitable. The competitive advantage comes from the quality of the relationship as experienced by the customer.

But it goes further than that.

Companies can no longer afford the cost of high staff turnover, either in terms of the expense of training or in terms of lost skills. As we move out of the labour-intensive world and pass through the technology-based work environment our staff are becoming more productive, better skilled, and more able to provide added value well beyond the cost of employment. But to obtain the benefit of this we must regard our people as our most valuable resource and must manage them very differently from the way we managed them in the past. The old style command and control approach is rapidly becoming inappropriate, a more collegiate, co-operative, participative leadership style is needed, and we *must* get our managers to change, to become the new leaders, if they and we are to survive.

As Peter Drucker says in his books *The New Realities* and *Post Capitalist Society*, we are moving into the age of the knowledge worker: our people are valuable because of the knowledge they have and how they apply it to gaining more knowledge. No longer are manual or even technical skills enough, our people want more than just to do things better – they want to develop new ways of doing things. Again, this is a management issue, but it is more than that – it is also an issue of retention of our most valuable resource, our people. If they feel constrained in their work, if they feel unable to deliver all the added value they are capable of, then they leave and their knowledge will go with them and we may well have lost our competitive advantage. Businesses are finding, more and more, that recruiting and retaining the right people is critical to their success and this means we must make our companies the ones that the best people want to work for. To do this we have to become a quality company – one in which quality is in everything we do, in every business process and in the way we manage. This is of paramount importance.

If quality in the relationship with the customer and in the way we run the company is so important, we cannot afford to ignore the necessity of change, of becoming a quality company. If we do ignore it, then our customers and our staff will leave to join our competitors and we will be out of business: and quicker than we think.

The question posed in the chapter title highlights current management thinking – we have achieved our goal so what do we do next? – but this is restrictive. We must look beyond the confines of the current business environment, beyond the current horizon, and we must make a step in faith. We must make survival our corporate goal, survival each day, each month, each year. And just as time rolls inexorably onwards so our fight for survival must go on and on. There is no ultimate destination on the quality journey. Each time we reach one peak there is one more peak beyond that which has to be climbed; a never-ending cycle of monitoring achievements against goals, re-defining the goals and obtaining new incremental improvements.

Where do we go next? The answer is staggeringly simple: back to the beginning. We have to revisit all our benchmarks and best practices, and we have to set new and more challenging goals that demand new incremental improvements in the way we do things. Achieving continuous incremental quality improvements must become a way of life: it is the only way our company will survive. CQI is not '**a** survival strategy' it is **THE survival strategy**.

Index

ABOUT THE AUTHOR:

Alasdair White is a consultant specialising in human resource management and organisational development (HRM & OD). Working for Performance Management Solutions Ltd, he is based near Brussels in Belgium and has an international practice with clients throughout Europe, the Middle East, and the Far East. Predominantly focused on the financial sector, his clients include international banks (such as Citibank, Union Bank of Switzerland, and Standard Chartered Bank), insurance companies, multinationals (such as IBM) and the European Commission. He is a visiting lecturer at the CPD Foundation in London and the London Management Centre at the University of Westminster.

With 10 years' experience in senior management and board level consultancy, Alasdair White now specialises in management development, training and quality programmes, and is a recognised authority on 'soft benchmarking' of HRM behaviours. He has considerable experience in, and knowledge of, multicultural and multi-lingual environments.

A frequent speaker on management development and much in demand as a seminar leader, he is the author of *Managing for Performance* (1995) and *Continuous Quality Improvement* (1996) both published by Piatkus Books of London.

Educated at King Alfred's College, Winchester, England, where he studied education and physical science, Alasdair White spent time teaching in Spain before becoming a business journalist and newspaper editor in the UK. He became a management consultant in 1984 and moved to The Netherlands in 1987. He has been in Belgium since 1992. He is an active member of the London-based Institute of Directors.

PERFORMANCE MANAGEMENT SOLUTIONS LTD

Performance Management Ltd is a human resource management (HRM) and organisational development (OD) consultancy with an international practice and reputation. With blue-chip clients in most of the countries in the European Union, as well as in the Middle East and Far East, the focus is mainly on the financial sector (large international banks) and on multi-nationals. However, since the PMS service is entirely transferable between sectors, the work they do applies to companies of all types and sizes.

As an HRM & OD consultancy, PMS specialises in three key areas:

- Management Development and Leadership
- Organisational Development
- Continuous Quality Improvement

Management Development and Leadership

Based firmly on the philosophy that the real development and future success of a business is based on how the company develops and manages its staff, PMS offers management training, leadership training, and management selection and assessment.

The key training programmes here are:

- **Performance Management Workshop** – a three to five day training course, for all levels of manager, which focuses on the key skills involved in managing people. Over 4000 managers worldwide have now participated in this course.

- **Performance Coaching Programme** – a two day specialised seminar designed to equip managers with coaching skills.

- **International Business Leadership Workshop** for those managers identified as having the potential for senior management and board level roles.

Each of these programmes is based on a set of core modules with optional extras – all of which are fully customised, in consultation with the client, to suit the company's culture and requirements.

Organisational Development

The work PMS does in this field is mainly in transition management, in which they develop and implement multicultural management development and cultural change programmes. In connection with this work they carry out major 'soft benchmarking' activities in which they audit and benchmark HRM behaviours through staff attitude surveys. This often leads to the development of internal communications strategies and the delivery of interpersonal communication skills and non-verbal skills training.

Continuous Quality Improvement

PMS believes that the continuous improvement in quality in all areas of a company is the only sure road to success and survival. Their work in CQI includes their other key areas of activity within an overall framework. The approach is to work closely with the client in identifying their requirements, developing the CQI Master Plan, and selecting those managers and other staff at all levels who will lead the process. PMS then takes a support and guidance role during the implementation phase and continuous feedback process. In this way the CQI programme becomes 'owned' by those involved throughout the company, thus enhancing the likelihood of success.

Working with clients

The PMS approach is based on a full understanding of the client's requirements, which is obtained through extensive interaction between the lead consultant and the client's senior management. This then leads to the development of the client-specific programme and to guiding the implementation. PMS's clients have found this approach to be most beneficial and the extended relationship with these companies confirms the value added through PMS's involvement.

To explore any of these areas in greater detail, please contact:

Performance Management Solutions Ltd
The Courtyard
12 Hill Street
St Helier
Jersey
Channel Islands

MANAGING FOR PERFORMANCE:
How to get the best out of yourself and your team

by Alasdair White
(Piatkus)

Performance management is crucial to a manager's success. To be an effective manager you need to concentrate on three main areas of responsibility: determining the objective of your team and how it will be reached; ensuring your team members are selected, developed and trained appropriately; and motivating the individuals so that the objective is achieved.

Managing For Performance is an accessible, practical guide to performance management techniques and how to apply them. It will appeal to all managers who want to improve their leadership skills and increase the efficiency of their team.

Techniques include:

- Understanding what makes people tick
- Choosing different management styles for different people
- Balancing goals and resources
- Convincing senior management that your goals are their goals
- Delegating wisely and productively
- Motivating your staff by creative coaching and training

Piatkus Business Books

Piatkus Business Books have been created for people who need expert knowledge readily available in a clear and easy-to-follow format. All the books are written by specialists in their field. They will help you improve your skills quickly and effortlessly in the workplace and on a personal level.

Titles include:

General Management and Business Skills

Be Your Own PR Expert: the complete guide to publicity and public relations Bill Penn
Complete Conference Organiser's Handbook, The Robin O'Connor
Complete Time Management System, The Christian H Godefroy and John Clark
Confident Decision Making J Edward Russo and Paul J H Schoemaker
Corporate Culture Charles Hampden-Turner
Energy Factor, The: how to motivate your workforce Art McNeil
Firing On All Cylinders: the quality management system for high-powered corporate performance Jim Clemmer with Barry Sheehy
How to Implement Change in Your Company John Spencer and Adrian Pruss
Influential Manager, The: How to develop a powerful management style Lee Bryce
Influential Woman, The: How to achieve success in your career – and still enjoy your personal life Lee Bryce
Leadership Skills for Every Manager Jim Clemmer and Art McNeil
Lure the Tiger Out of the Mountains: timeless tactics from the East for today's successful manager Gao Yuan
Managing For Performance Alasdair White
Managing Your Team John Spencer and Adrian Pruss
Outstanding Negotiator, The Christian H Godefroy and Luis Robert
Problem Solving Techniques That Really Work Malcolm Bird
Right Brain Time Manager, The Dr Harry Alder
Seven Cultures of Capitalism, The: value systems for creating wealth in Britain, the United States, Germany, France, Japan, Sweden and the Netherlands Charles Hampden-Turner and Fons Trompenaars
Smart Questions for Successful Managers Dorothy Leeds
Strategy of Meetings, The George David Kieffer

Think Like A Leader Dr Harry Alder

Personnel and People Skills

Best Person for the Job, The Malcolm Bird
Dealing with Difficult People Roberta Cava
Problem Employees: how to improve their behaviour and their performance Peter Wylie and Mardy Grothe
Psychological Testing for Managers Dr Stephanie Jones
Tao of Negotiation: How to resolve conflict in all areas of your life Joel Edelman and Mary Beth Crain

Financial Planning

Better Money Management Marie Jennings
Great Boom Ahead, The Harry Dent
How to Choose Stockmarket Winners Raymond Caley
Perfectly Legal Tax Loopholes Stephen Courtney
Practical Fundraising For Individuals And Small Groups David Wragg

Small Business

50 Businesses to Start from Home Mel Lewis
How to Earn Money from Your Personal Computer Polly Bird
How to Run a Part-Time Business Barrie Hawkins
Making Money From Your Home Hazel Evans
Making Profits: a six-month plan for the small business Malcolm Bird
Marketing On A Tight Budget Patrick Forsyth
Profit Through the Post: How to set up and run a successful mail order business Alison Cork

Motivational

Play to Your Strengths Donald O Clifton and Paula Nelson
Super Success Philip Holden
Winning Edge, The Charles Templeton

Self-Improvement

Brain Power: the 12-week mental training programme Marilyn vos Savant and Leonore Fleischer
Create the Life You Want With NLP Karen Holding
Creating Abundance Andrew Ferguson
Creative Thinking Michael LeBoeuf
Getting What You Want Quentin de la Bedoyere
Memory Booster: easy techniques for rapid learning and a better memory Robert W Finkel
Napoleon Hill's Keys To Success Matthew Sartwell (ed.)

Napoleon Hill's Unlimited Success Matthew Sartwell (ed.)
NLP: The New Art and Science of Getting What You Want Dr Harry Alder
Organise Yourself Ronni Eisenberg with Kate Kelly
Personal Growth Handbook, The Liz Hodgkinson
Personal Power Philippa Davies
Quantum Learning: unleash the genius within you Bobbi DePorter with Mike Hernacki
Right Brain Manager, The: how to use the power of your mind to achieve personal and professional success Dr Harry Alder
10-Minute Time And Stress Management Dr David Lewis
Three Minute Mediator, The David Harp with Nina Feldman
Total Confidence Philippa Davies

Sales and Customer Services
Art of the Hard Sell, The Robert L Shook
Commonsense Marketing For Non-Marketers Alison Baverstock
Creating Customers David H Bangs
Guerrilla Marketing Excellence Jay Conrad Levinson
Guerrilla Marketing Jay Conrad Levinson
Guerrilla Marketing On The Internet Jay Conrad Levinson and Charles Rubin
How to Close Every Sale Joe Girard
How to Make Your Fortune Through Network Marketing John Bremner
How to Succeed in Network Marketing Leonard Hawkins
How to Win a Lot More Business in a Lot Less Time Michael LeBoeuf
How to Win Customers and Keep Them for Life Michael LeBoeuf
How to Write Letters that Sell Christian Godefroy and Dominique Glocheux
One-To-One Future, The Don Peppers and Martha Rogers
Sales Power: the Silva mind method for sales professionals José Silva and Ed Bernd Jr
Selling Edge, The Patrick Forsyth
Telephone Selling Techniques That Really Work Bill Good
Winning New Business: a practical guide to successful sales presentations Dr David Lewis

Presentation and Communication
Better Business Writing Maryann V Piotrowski
Complete Book of Business Etiquette, The Lynne Brennan and David Block

Confident Conversation Dr Lillian Glass
Confident Speaking: how to communicate effectively using the Power Talk System Christian H Godefroy and Stephanie Barrat
He Says, She Says: closing the communication gap between the sexes Dr Lillian Glass
Personal Power Philippa Davies
Powerspeak: the complete guide to public speaking and presentation Dorothy Leeds
Presenting Yourself: a personal image guide for men Mary Spillane
Presenting Yourself: a personal image guide for women Mary Spillane
Say What You Mean and Get What You Want George R. Walther
Your Total Image Philippa Davies

Careers and Training
How to Find the Perfect Job Tom Jackson
Jobs For The Over 50s Linda Greenbury
Making It As A Radio Or TV Presenter Peter Baker
Marketing Yourself: how to sell yourself and get the jobs you've always wanted Dorothy Leeds
Networking and Mentoring: a woman's guide Dr Lily M Segerman-Peck
Perfect CV, The Tom Jackson
Perfect Job Search Strategies Tom Jackson
Secrets of Successful Interviews Dorothy Leeds
Sharkproof: get the job you want, keep the job you love in today's tough job market Harvey Mackay
10-Day MBA, The Steven Silbiger
Ten Steps To the Top Marie Jennings
Which Way Now? – how to plan and develop a successful career Bridget Wright

For a free brochure with further information on our complete range of business titles, please write to:

Piatkus Books
Freepost 7 (WD 4505)
London W1E 4EZ